# Lecture Notes in Networks and Systems

Volume 352

The series "Lecture Notes in Networks and Systems" publishes the latest developments in Networks and Systems—quickly, informally and with high quality. Original research reported in proceedings and post-proceedings represents the core of LNNS.

Volumes published in LNNS embrace all aspects and subfields of, as well as new challenges in, Networks and Systems.

The series contains proceedings and edited volumes in systems and networks, spanning the areas of Cyber-Physical Systems, Autonomous Systems, Sensor Networks, Control Systems, Energy Systems, Automotive Systems, Biological Systems, Vehicular Networking and Connected Vehicles, Aerospace Systems, Automation, Manufacturing, Smart Grids, Nonlinear Systems, Power Systems, Robotics, Social Systems, Economic Systems and other. Of particular value to both the contributors and the readership are the short publication timeframe and the world-wide distribution and exposure which enable both a wide and rapid dissemination of research output.

The series covers the theory, applications, and perspectives on the state of the art and future developments relevant to systems and networks, decision making, control, complex processes and related areas, as embedded in the fields of interdisciplinary and applied sciences, engineering, computer science, physics, economics, social, and life sciences, as well as the paradigms and methodologies behind them.

Indexed by SCOPUS, INSPEC, WTI Frankfurt eG, zbMATH, SCImago.

All books published in the series are submitted for consideration in Web of Science.

More information about this series at http://www.springer.com/series/15179

Grzegorz Sierpiński
Editor

# Intelligent Solutions for Cities and Mobility of the Future

Springer

*Editor*
Grzegorz Sierpiński
Department of Transport Systems,
Traffic Engineering and Logistics
Silesian University of Technology
Gliwice, Poland

ISSN 2367-3370 ISSN 2367-3389 (electronic)
Lecture Notes in Networks and Systems
ISBN 978-3-030-91155-3 ISBN 978-3-030-91156-0 (eBook)
https://doi.org/10.1007/978-3-030-91156-0



This Springer imprint is published by the registered company Springer Nature Switzerland AG
The registered company address is: Gewerbestrasse 11, 6330 Cham, Switzerland

# Preface

Solutions addressed to cities and their inhabitants should maximize the quality of life while reducing the negative impact on the environment. Technological and spatial development is constantly bringing new opportunities, also in the field of mobility. Identification of real needs requires the use of modern solutions in the field of acquiring as well as advanced tools for data analysis. This knowledge can then be used to improve the planning and organization of traffic flows.

This publication contains selected papers submitted to and presented at the 17th "Transport Systems. Theory and Practice" Scientific and Technical Conference organized by the Department of Transport Systems, Traffic Engineering and Logistics at the Faculty of Transport and Aviation Engineering of the Silesian University of Technology (Katowice, Poland). The problems addressed in the publication entitled *Intelligent Solutions for Cities and Mobility of the Future* have been divided into four parts:

- Part 1. Usefulness of new forms of mobility,
- Part 2. Use of ITS data in planning and organization of traffic flows,
- Part 3. Intelligent and integrated transport systems
- Part 4. Improving the efficiency of ecological friendly solutions in transport systems.

The articles included in the publication are expressions of case study-based scientific and practical approach to the problems of contemporary transport systems. The authors drew attention to changes in transport processes resulting from many aspects, including technological development, shaping the needs of travelers or a sudden reduction in travel caused by the pandemic phenomenon. For all these situations, there is a need to use modern solutions, often requiring high dynamics in traffic planning and management. The publication also includes an abundance of numerical analyses based on significant data sets, illustrating the close affiliation between intelligent solutions and future mobility.

I would like to express my deepest gratitude to all authors, for reflecting the key problems of contemporary transport systems in a concise manner, as well as to reviewers, in recognition of their insightful remarks and suggestions without which this collection of papers would have never been published.

September 2021 Grzegorz Sierpiński

# Organization

17th Scientific and Technical Conference "Transport Systems. Theory and Practice" (TSTP2021) is organized by the Department of Transport Systems, Traffic Engineering and Logistics, Faculty of Transport and Aviation Engineering, Silesian University of Technology (Poland).

## Organizing Committee

### Organizing Chair

| | |
|---|---|
| Grzegorz Sierpiński | Silesian University of Technology, Poland |

### Members

| | |
|---|---|
| Grzegorz Karoń | Silesian University of Technology, Poland |
| Marcin Staniek | Silesian University of Technology, Poland |
| Ireneusz Celiński | Silesian University of Technology, Poland |
| Marcin J. Kłos | Silesian University of Technology, Poland |
| Barbara Borówka | Silesian University of Technology, Poland |
| Adrian Barchański | Silesian University of Technology, Poland |
| Agata Kurek | Silesian University of Technology, Poland |
| Piotr Soczówka | Silesian University of Technology, Poland |

## The Conference took place under the Honorary Patronage

Ministry of Infrastructure
Marshal of the Silesian Voivodeship
Committee on Transport of Polish Academy of Sciences

## Scientific Committee

Elżbieta Załoga University of Szczecin, Poland
Stanisława Zamkowska Blessed Priest Władysław Findysz University of Podkarpacie in Jaslo
Jacek Żak Poznań University of Technology
Jolanta Żak Warsaw University of Technology, Poland
Renata Żochowska Silesian University of Technology, Poland

## Referees

Ireneusz Celiński
Jacek Chmielewski
Piotr Czech
Boris Davydov
Elżbieta Macioszek
Grzegorz Sierpiński
Marcin Staniek
Agnieszka Tubis
Grzegorz Wojnar

# Contents

# Usefulness of New Forms of Mobility

# Information and Usage Asymmetry of Shared Mobility Services Among Different Generations

Andrzej Hanusik(✉)

University of Economics in Katowice, Katowice, Poland
andrzej.hanusik@ue.katowice.pl

**Abstract.** The purpose of this paper is to examine the extent of information and usage asymmetry among representatives of different generations regarding shared mobility services. This paper is based on the literature review and survey that was carried out on people belonging to different generations (Baby Boomers and Generations X, Y, Z). The conducted research confirmed the hypothesis that younger people have more information about shared mobility services and more often use these solutions in everyday life. Differences in asymmetry were also observed - the most popular concepts (carsharing, bikesharing, scootersharing, carpooling) are characterized by left-handed asymmetry, while the less popular concepts (vanpooling, ridesharing, ride - hailing) are right-sided. The main limitation of the study was the origin of the respondents from only one urban area, the snowball sampling method and use of the English nomenclature.

**Keywords:** Sharing economy · Information asymmetry · Shared mobility services · Generations

## 1 Introduction

The sharing economy is undeniably a phenomenon that has largely entered to the contemporary urban mobility. Despite the huge popularity of this concept, there is still no universally accepted definition, which may be caused by its relative novelty [1]. It should be noted, that manifestations of sharing in various forms took place already in prehistoric times [2], so this phenomenon is new only from a scientific perspective. Shared mobility solutions are one of the categories directly related to the sharing economy. They allow users for a short term access to means of transport if the transport need appears. The most popular shared mobility solutions include carsharing, bikesharing and carpooling [3].

Each concept corresponding to the assumptions of the sharing economy (including shared mobility) creates very complex network structures [4]. For the efficient functioning of any shared solution, the participation of many entities is necessary (suppliers, consumers, people responsible for the functioning of a given platform and other stakeholders, such as non-governmental organizations or public administration bodies). Therefore sharing economy is based on peer-to-peer networks [5]. It should be noted


G. Sierpiński (Ed.): TSTP 2021, LNNS 352, pp. 3–13, 2022.
https://doi.org/10.1007/978-3-030-91156-0_1

that the globalization processes, that are characteristic for the modern world, have contributed to the market changes - now they are characterized not only by a local scope of influence, but also in the global scale [6]. That situation contributed to the extension of the scope of offered activities in the field of the sharing economy. Specialized entities offering shared services (which previously operated mainly within small social groups) were also established. Moreover, entities participating in shared relationships also differ. Large number and diversification of entities involved in sharing, means that individual entities have different information, both in terms of shared mobility services quality and quantity - the problem of information asymmetry appears, which was first described by George A. Akerlof [7]. Additionally, the problem of usage asymmetry of these solutions by people belonging to different generations appears. It can be assumed that younger people are more familiar with shared mobility services than older people.

In modern world there are many concepts that correspond to the assumptions of sharing economy. In case of shared mobility they may be limited to seven solutions:

- Carsharing – access to vehicles based on one time or periodic payments [8],
- Bikesharing – a city bike system [9] that allows you to rent a bicycle based on one time or periodic payments (similar to carsharing),
- Scootersharing – access to scooters based on one time or periodic payments. It can apply to both electric and combustion scooters [10],
- Carpooling – sharing a ride with a driver and one or more passengers [11],
- Vanpooling – a way to travel in a larger vehicle, such as a bus or van, where people are grouped into clusters based on their place of residence. Such journeys are made mainly as part of travel to work [12],
- Ridesharing – a way of traveling in which individual travelers share the vehicle and travel costs (fuel, tolls, parking fees) with other people with a similar planned route and travel time [13],
- Ride - hailing – a platform that allows to immediately summon a car and a driver to come and take a passenger to a choosen destination. Ride - hailing platforms do not have to possess their own fleet or hire drivers as employees [14].

However, the scale of asymmetry in concepts related to sharing economy remains unsolved. There are also voices that the sharing economy by its very assumptions can counteract the phenomenon of information asymmetry [15]. It should be noted, that very often for the functioning of shared mobility, it is necessary to have a transaction platform - it may be a different type of product, service or enterprise witch mediates in transactions between two or more entities [16]. Such situation makes it necessary to have appropriate knowledge and information for effective participation in sharing. It can be assumed that younger people have more information about shared mobility than older ones. The aim of this paper is to determine the scale of shared mobility services information and usage asymmetry among people belonging to different generations, because older people usually have less knowledge about modern concepts than younger generations. In order to verify this hypothesis, a questionnaire survey was carried out on people belonging to different generations that was a base for determination the scale of the information and usage asymmetry of shared mobility services.

The article consists of four main parts. In the first part, the research issues were introduced and the research gap was identified. The second part presents the methodology used in the article. In the third part the obtained research results was showed. The fourth part consists of conclusions and discussion about the article content.

## 2 Methodology and Data

The article was based on surveys conducted on representatives of various generations. An arbitrary selection of the sample was used, because the aim of the study is to obtain a new view on the presented problem, so the cross-section of opinions on this subject for the entire population is not important (the sample does not have to be representative) [17].

The survey was conducted between 17 and 31 August 2020 on a group of 584 respondents. The survey was carried out both in the traditional, paper form and via Internet. Such decision was related to the large age diversification of the surveyed group - younger people prefer electronic questionnaires, while older people very often cannot use IT solutions. It should also be noted that due to the epidemiological situation (COVID-19 pandemic), collecting materials (especially among representatives of older generations) was significantly more difficult, which was directly translated into the structure of received responses. In order to reach a larger group of respondents the snowball method was used - questionnaires were distributed to a group of people who were asked to provide the form to people of the same age and gender [18]. This approach made it possible to reach a larger group of respondents. It should also be noted that the respondents came mainly from the area of the Upper Silesia – Zagłębie Metropolis (GZM), which can be considered as a limitations of the conducted research.

The survey consisted of 7 main questions, which had the form of a test, short written answer, linear scale or Libert's scale, and a demographic section, which allowed the answers to be assigned to defined age groups - Silent Generation, Baby Boomers, X generation, Y generation (Millenials) and the post-Millenial generation [19] (generation Z and Alpha) - however, there is no one rigid, universally accepted age division allowing for an unambiguous assignment of a particular person to a given generation. Therefore, the author made the following assumptions:

- age from 0 to 23 years - post–Millennials generation (Z and Alpha),
- age from 24 to 40 – Y generation (Millenials),
- age from 41 to 55 – X generation,
- age from 56 to 73 – Baby Boomers,
- age 74 and over – Silent Generation.

After obtaining the answers from the respondents, statistical analysis was conducted. The first step was to convert the responses (from qualitative to quantitative). Most of the questions in the survey were based on the Likert scale permutation, which allowed the respondents to determine which of the answers best describe their knowledge of a given concept of shared mobility services. For the purposes of the analysis, numerical values were assigned to individual categories of responses, in accordance with the assumption:

$$V_i \ \varepsilon < 1; k > \wedge \ k \ \varepsilon \ \mathrm{N} \tag{1}$$

where:

- $V_i$ – assigned value for a given response category,
- k – number of categories,
- N – natural number.

Then, the number of individual categories of answers for particular generations was presented in tabular form. The average value of the indicator for each generation was calculated using the given index:

$$aV_i = \frac{\sum_{i=1}^{n}(x_i V_i)}{\sum_{i=1}^{n} x_i} \tag{2}$$

where:

- $aV_i$ - the average value of the indicator for a given observation and age group,
- $x_i$ - the number of indications for a given response category.

For each shared mobility service, the value of the asymmetry coefficient was also calculated, along with the information whether the distribution of the obtained responses is characterized by left or right asymmetry. For this purpose, the classic asymmetry coefficient index based on the third moment of central tendency was used. This indicator has the form [20]:

$$A_j = \frac{\sqrt{n(n-1)}}{n-2} \times \frac{\frac{1}{n}\sum_{i=1}^{n}(\mathrm{aVi} - m\mathrm{aVi})^3}{\left(\frac{1}{n}\sum_{i=1}^{n}(\mathrm{aVi} - m\mathrm{aVi})^2\right)^{3/2}} \tag{3}$$

where:

- $A_j$ – the asymmetry coefficient factor for a given concept,
- n – number of observations,
- $maV_i$ – average of the an average value of the indicator for a given observation and age group.

The obtained factor determines whether the greater part of the studied population has values of features below the mean (right-sided asymmetry) or above it (left-sided asymmetry). This information allows to draw a conclusion about the respondents' awareness of a given concept - more popular solutions, which are more often visible on city streets and present in the media, should be characterized by left-sided asymmetry.

The last element of the study was to determine the synthetic indicator of all concepts in the field of the sharing economy for representatives of individual generations, which will ultimately verify or falsify the posed hypothesis:

$$sV = \frac{\sum_{i=1}^{n} \mathrm{aV}_i}{n} \tag{4}$$

The obtained values were used to create a model of awareness of the concept of shared mobility among different generations. Based on the model, the nature of the relationship between belonging to a given generation and knowledge of the concept of shared mobility can be determined.

This structure of the research procedure allowed to obtain the necessary data and to make a verification or falsification of the research hypothesis, which will fulfill the main goal of the research project.

## 3 Results

Figure 1 shows the structure of respondents, taking into account their gender and age (as well as their belonging to particular generations in accordance with the division proposed in the previous section).

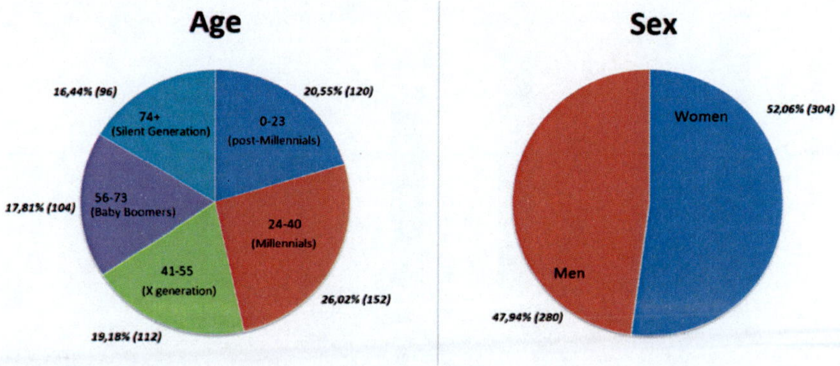


**Fig. 1.** Details of the respondents

The first question concerned respondents' knowledge about broadly understood concept of the sharing economy. Respondents stated whether they knew such a concept or had only heard about it without any details or whether they were completely unfamiliar with such solutions. Table 1 shows the obtained results. The average value of the indicator was highlighted with grey colour.

The post-Millennials and the X generations are most aware of the concept of sharing economy. The Millennial generation also achieved a high result. The oldest people, who belong to the Silent Generation and the Baby Boomers have the lowest awareness of the concept of sharing economy. The hypothesis about the existence of asymmetry between different generations has also been confirmed - it can be assumed that younger people have a wide range of information about the sharing economy. A biggest difference has been identified between Generation X and Baby Boomers. It should also be noted that the coefficient of asymmetry is $-1.21$, which indicates a left-asymmetric distribution - the greater part of the studied sample has values of the features above the average.

The main part of the survey focused on the different shared mobility concepts. Figure 2 shows the most recognizable concepts in the field of shared mobility.

The most recognizable concepts in the field of shared mobility are bikesharing, carsharing, scootersharing and carpooling. The least recognizable concepts are vanpooling,

**Table 1.** Awareness of the concept of the sharing economy

| | Number of responses according to the age criterion | | | | |
|---|---|---|---|---|---|
| | 0–23 | 24–40 | 41–55 | 56–73 | 74+ |
| 1 - I have never heard of such concept | 13 | 21 | 3 | 27 | 66 |
| 2 - I have heard of such concept, but I do not know details | 11 | 18 | 29 | 53 | 30 |
| 3 - I know this concept | 96 | 113 | 80 | 24 | 0 |
| **Average value** | **2,69** | **2,61** | **2,69** | **1,97** | **1,31** |

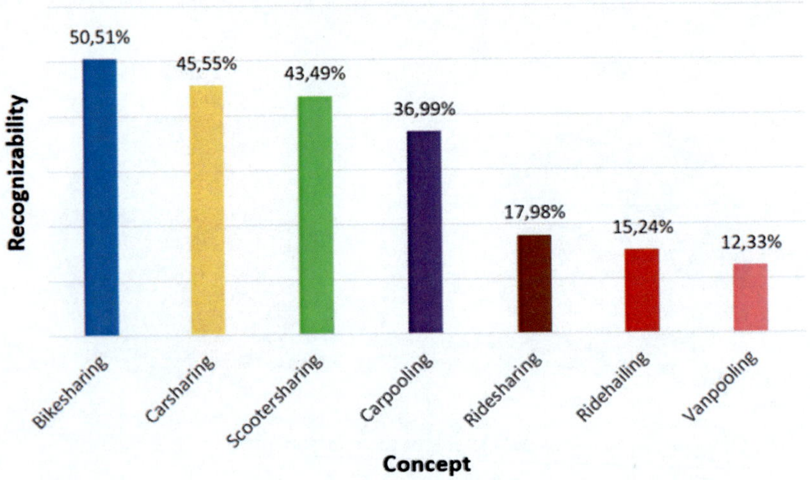


**Fig. 2.** Most recognizable concepts in the field of shared mobility

ridehailing and ridsharing. Interestingly, very often people who declared that they used ridehailing platforms (e.g. Uber or Bolt) or ridesharing (e.g. blablacar) did not know the concepts themselves. Table 2 shows the number of responses in each category and the average value of the indicator (highlighted with grey colour). The following gradation of the response was adopted:

- 1 - I have never heard of such concept,
- 2 - I have heard about such concept, but I do not know its meaning,
- 3 - I know this concept, but I have never used it,
- 4 - I know this concept and I have used it once,
- 5 - I know this concept and use it regularly.

The hypothesis that younger people have more information about shared mobility services and use it more often has been confirmed again. In the case of carsharing, bikesharing and scootersharing, the highest results were obtained by people belonging to the Millennial generation. It can also be noticed that the vanpooling, ridesharing and

**Table 2.** Shared mobility across different generations

| Concept | Value | Number of responses according to the age criterion | | | | |
|---|---|---|---|---|---|---|
| | | 0–23 | 24–40 | 41–55 | 56–73 | 74+ |
| Carsharing | 1 | 5 | 14 | 18 | 32 | 83 |
| | 2 | 67 | 18 | 17 | 51 | 13 |
| | 3 | 7 | 65 | 70 | 21 | 0 |
| | 4 | 41 | 23 | 7 | 0 | 0 |
| | 5 | 0 | 32 | 0 | 0 | 0 |
| **Average value** | | **2,70** | **3,27** | **2,59** | **1,89** | **1,14** |
| Bikesharing | 1 | 11 | 15 | 17 | 37 | 90 |
| | 2 | 50 | 9 | 9 | 45 | 6 |
| | 3 | 19 | 55 | 72 | 22 | 0 |
| | 4 | 40 | 16 | 6 | 0 | 0 |
| | 5 | 0 | 57 | 8 | 0 | 0 |
| **Average value** | | **2,73** | **3,27** | **2,59** | **1,89** | **1,14** |
| Scootersharing | 1 | 22 | 24 | 33 | 64 | 87 |
| | 2 | 58 | 8 | 9 | 16 | 9 |
| | 3 | 7 | 88 | 64 | 24 | 0 |
| | 4 | 33 | 16 | 6 | 0 | 0 |
| | 5 | 0 | 16 | 0 | 0 | 0 |
| **Average value** | | **2,43** | **2,95** | **2,39** | **1,62** | **1,09** |
| Carpooling | 1 | 32 | 32 | 72 | 56 | 88 |
| | 2 | 8 | 40 | 8 | 24 | 8 |
| | 3 | 24 | 40 | 13 | 16 | 0 |
| | 4 | 56 | 24 | 10 | 8 | 0 |
| | 5 | 0 | 16 | 9 | 0 | 0 |
| **Average value** | | **2,87** | **2,68** | **1,89** | **1,78** | **1,08** |
| Vanpooling | 1 | 59 | 96 | 99 | 97 | 96 |
| | 2 | 21 | 24 | 13 | 7 | 0 |
| | 3 | 30 | 32 | 0 | 0 | 0 |
| | 4 | 10 | 0 | 0 | 0 | 0 |
| | 5 | 0 | 0 | 0 | 0 | 0 |
| **Average value** | | **1,93** | **1,58** | **1,17** | **1,07** | **1,00** |

*(continued)*

**Table 2.** (*continued*)

| Concept | Value | Number of responses according to the age criterion | | | | |
|---|---|---|---|---|---|---|
| | | 0–23 | 24–40 | 41–55 | 56–73 | 74+ |
| Ridesharing | 1 | 55 | 113 | 95 | 95 | 96 |
| | 2 | 9 | 7 | 9 | 0 | 0 |
| | 3 | 16 | 25 | 8 | 9 | 0 |
| | 4 | 9 | 0 | 0 | 0 | 0 |
| | 5 | 31 | 7 | 0 | 0 | 0 |
| **Average value** | | **2,60** | **1,56** | **1,22** | **1,17** | **1,00** |
| Ride - hailing | 1 | 55 | 121 | 88 | 104 | 96 |
| | 2 | 16 | 7 | 8 | 0 | 0 |
| | 3 | 7 | 16 | 11 | 0 | 0 |
| | 4 | 9 | 0 | 5 | 0 | 0 |
| | 5 | 33 | 8 | 0 | 0 | 0 |
| **Average value** | | **2,58** | **1,47** | **1,40** | **1,00** | **1,00** |

ride – hailing have a lower average score – in these concepts, people belonging to the post – Millennial generation are the most informed. Calculations were also made to determine the asymmetry coefficient index – the results are presented in Table 3.

**Table 3.** Asymmetry coefficient index of shared mobility services

| Concept | Asymmetry coefficient | Asymmetry |
|---|---|---|
| Carsharing | −0,58 | Left-sided |
| Bikesharing | −0,60 | Left-sided |
| Scootersharing | −0,46 | Left-sided |
| Carpooling | −0,22 | Left-sided |
| Vanpooling | 0,93 | Right-sided |
| Ridesharing | 1,72 | Right-sided |
| Ride - hailing | 1,63 | Right-sided |

In the case of bikesharing, carsharing, scootersharing and carpooling the greater part of the analyzed sample is above the average level of the examined feature. It should be noted that these are the most popular concepts (see Fig. 2). For less popular concepts (ridesharing, ride - hailing and vanpooling), the greater part of the analyzed sample is below the average level of the examined feature. Thus, it can be stated that the more popular concepts are characterized by left-sided asymmetry, and less popular ones are characterized by right-sided asymmetry.

The last part of the study was to determine the average value for all concepts in the field of shared mobility services (average of averages). This procedure allowed to obtain a single value that characterizes the entire generation, thanks to which it will be possible to clearly define the information and usage asymmetry among different generations. The results are shown in Fig. 3.

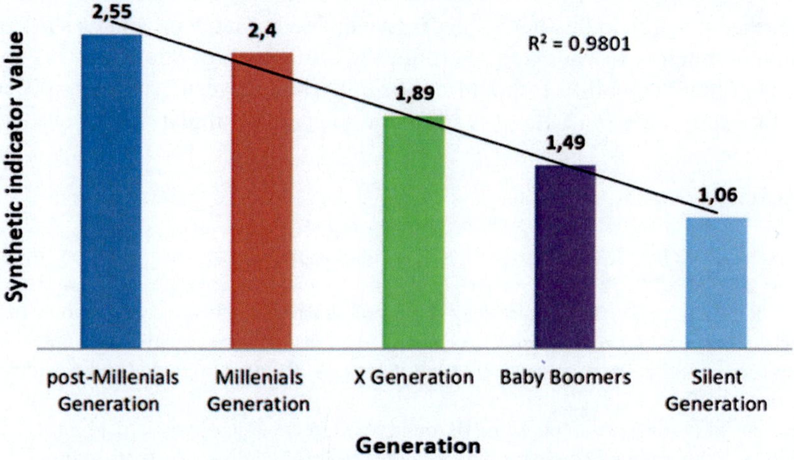


**Fig. 3.** Information and usage asymmetry among different generations

The results clearly confirm that there is a correlation between belonging to a given generation and the awareness of shared mobility services. The biggest difference can be observed between Millennials Generation and X Generation. This may be due to the fact that the Millennials used technology practically all their lives. Also it should be noted that the difference between Millennials and post-Millennials generation is insignificant. This situation allows to define the main target group for shared mobility services - Millennials and post-Millennials Generation.

## 4 Conclusions and Discussion

The presented article is a new look at the information and usage asymmetry in shared mobility services. According to the author's knowledge, there are no studies dealing with similar topics in the literature so far. Thus, the article may become an impulse for an extended academic debate and form the basis for further research.

A relation between the awareness and usage of the concept of shared mobility and belonging to a given generation was observed. The post-Millennials and the X generations are most aware of the sharing economy concept (Millennials are also very familiar to it). The oldest people (Silent Generation and the so-called Baby Boomers) have the lowest awareness of the concept of the sharing economy. It can also be observed that concepts that appear less frequently in the media have a lower average score. The main limitation of the study was the origin of the respondents from only one urban centre and the snowball sampling method. The English nomenclature of shared mobility that was used in the article can also be considered as a limitation - this situation could have contributed to the partial exclusion of people who do not speak English.

The conducted research made it possible to achieve the main goal of the article, i.e. to determine the scale of information asymmetry in the concept of shared mobility. It should be noted that the collected data come from people living in the Upper Silesia – Zagłębie Metropolis (GZM), so the obtained results cannot be considered typical for all areas. Further research in this field is necessary in another large urban areas in Poland - only their compilation will allow to determine the actual scale of information asymmetry in the field of shared mobility. It should also be noted that the results may also differ from country to country, which is why it is advisable to conduct similar studies elsewhere.

## References

1. Frenken, K., Schor, J.: Putting the sharing economy into perspective. Environ. Innov. Soc. Trans. **23**, 3–10 (2017)
2. Belk, R., Llamas, R.: The nature and effects of sharing in consumer behaviour. In: Mick, D. G., Pettigrew, S., Pechmann, C., Ozanne, J.L. (eds.) Transformative consumer research for personal and collective well-being, pp. 625–646, Routledge Taylor and Francis Group, London (2012)
3. Shaheen, S., Cohen, A., Zohdy, I., Berman, W., Greenberg, A.: Shared Mobility: Current Practices and Guiding Principles, United States Federal Highway Administration (2016)
4. Hanusik, A.: Wpływ ekonomii współdzielenia na zmianę linearnego modelu biznesowego w kierunku ujęcia sieciowego. In: Antonowicz, P., Chmielewski, M., Pisarewicz, P. (eds.) Ekonomia i prawo. Wybrane zagadnienia z zakresu zarządzania rozwojem przedsiębiorstwa, pp. 41–55, Wydawnictwo Uniwersytetu Gdańskiego, Gdańsk (2019)
5. Redfearn III, R.L.: Sharing economy misclassification: employees and independent contractors in transportation network companies. Berkeley Technol. Law J. **31**(2), 1022–1056 (2016)
6. Malara, Z., Rzęchowski, J.: Zarządzanie informacją na rynku globalnym. Teoria i praktyka, Wydawnictwo C. H. Beck, Wrocław (2011)
7. Akerlof, G.A.: The market for "Lemons": quality uncertainty and the market mechanism. Q. J. Econ. **84**(3), 488–500 (1970)
8. Bardhi, F., Eckhardt, G.M.: Access-based consumption: the case of car sharing. J. Consum. Res. **39**(4), 881–898 (2012)
9. DeMaio, P.: Bike-sharing: history, impacts, models of provision, and future. J. Public Transp. **12**(4), 41–56 (2009)
10. Howe, E.: Globa Scootersharing Market Report, Innovationszentrum für Mobilität und gesellschaftlichen Wandel (2018)
11. Kelly, K.L.: Casual carpooling-enhanced. J. Public Transp. **10**(4), 119–130 (2007)
12. Pratsch, L., Starling, R.: Vanpooling. an Update, U. S. Department of Energy (1979)
13. Furuhata, M., Dessouky, M., Ordóñez, F., Brunet, M., Wang, X., Koenig, S.: Ridesharing: The state-of-the-art and future directions. Transp. Res. Part B Methodol. **57**, 28–46 (2013)
14. Young, M., Farber, S.: The who, why, and when of Uber and other ride-hailing trips: an examination of a large sample household travel survey. Transp. Res. Part A Policy Pract. **119**, 383–392 (2019)
15. Thierer, A., Koopman, C., Hobson, A., Kuiper, C.: How the Internet, the sharing economy, and reputational feedback mechanisms solve the lemons problem. Univ. Miami Law Rev. **70**(3), 830–878 (2016)

16. Rochet, J., Tirole, J.: Platforms competition in two-sided markets. J. Eur. Econ. Assoc. **1**(4), 990–1029 (2003)
17. Churchill, G.A.: Badania marketingowe, Wydawnictwo Naukowe PWN, Warszawa (2002)
18. Goodman, L.A.: Snowball sampling. Ann. Math. Stat. **32**(1), 148–170 (1961)
19. Fry, R.: Millennials Overtake Baby Boomers as America's Largest Generation, Pew Research Center (2016)
20. Doane, D.P., Seward, L.E.: Measuring skewness: a forgotten statistic? J. Statist. Educ. **19**(2) (2011)

# Measuring and Forecasting of Passengers Modal Split Through Road Accidents Statistical Data

Ahmed Jaber(✉) and János Juhász

Department of Transport Technology and Economics,
Budapest University of Technology and Economics, Stoczek u. 2, Budapest 1111, Hungary
ahjaber6@edu.bme.hu, juhasz.janos@kjk.bme.hu

**Abstract.** This research investigates the modal split of passengers' transport in a new approach different from the dominant methodology of four-step demand model, using accidents data in calculation, assuming that the more accidents for a mode (cars, public transport, and bicycles), the more passengers who are using it. More than 100,000 road accidents are analyzed for consecutive four years in Hungary. Drivers, front-seat and rear-seat passengers are categorized through age intervals. Age distribution profiles for each mode of private cars, public transport, cycling, motors are developed after normalized the intervals with their corresponding population. Hypothesis of using the approach of finding passenger modal split through road accident data is tested as overall percentages in macro level as Hungary scale and selected cities in micro level for cities of Budapest, Miskolc, Pecs, and Szeged. Forecasting of the passenger cars split ratio is developed for four years ahead depending on time-series decomposition method. Results show that modes of passenger cars, and walking and cycling percentages are acceptable with limitations in reality, unlike public transport which is rejected at all times to be estimated through road accidents data, while forecasting passenger cars is convenient.

**Keywords:** Modal split · Road accidents · Passengers transport · Public transport · Passengers' safety · Forecasting

## 1 Introduction

The modal split is commonly used in traffic engineering. As a short definition, it is the proportion of each transportation mode in a town, city, region, or a country. The simple method is to count the volumes of each mode used in passenger trips or freight transport, and compare them to each other in order to receive valuable ratios which are described as the modal split. Other researchers call it as modal share. These ratios vary from city to city, or other micro and macro scales, and is applied to evaluate existing conditions or development in transportation [1]. Historically, the first mention to this term was in the United States in the 1950s, as transportation became a serious issue there. Several theories about traffic forecasting and traffic models which tried to predict traffic volumes in future. The very classic transportation demand model used


G. Sierpiński (Ed.): TSTP 2021, LNNS 352, pp. 14–26, 2022.
https://doi.org/10.1007/978-3-030-91156-0_2

there was the four step model. The third stage of modelling is called modal split [2]. The modal split in transport is directly linked to the idea that modal speed, transport times, transport management and localizations determine modal shares [3]. Modal split is a crucial indicator for evaluating the sustainability of urban public transport systems [4]. Studying mobility behavior universally, it is found that on average all persons consume about one to one and a half hour per day for their mobility and make about three to four trips per day – no matter where they live or what their cultural background is. The only significant difference is the transport mode they use for their mobility [5]. This rises the importance of investigating the mode share ratios among different models which lead to a better mobility management for individuals, cities, and economies.

Calculating modal split as a part of aggregation of four step model leads to some limitations in this method, that is because this method does not consider individuals' differences, but only proprieties with zones (e.g., average travel time between two zones) [6]. In the traditional technique, researchers have developed models based on population divided into groups with the same behavior in the transport process. They are called disaggregated or individual models. The major drawbacks are that they require for exact computing technology and detailed data. The essential way to obtain the necessary data is conducting a survey, which provides the best quality information of transport relations if surveys conducted directly at home, well known as traffic sociological survey [7]. On the other hand, surveys or questionnaires have several negative sides such as: respondents may provide dishonest answers, leave questions as blanks, not understand the requirements, and/or not feel comfortable providing answers. In addition, this process is time consuming and may produce a non-representative sample.

Therefore, this research finds new methodology to investigate the mode split for passengers in a macro scale as a country (Hungary) or larger scale as micro level of cities. This research is using accidents data to calculate modal split, assuming that the more accidents for a mode, the more passengers who are using it. Several researchers have emphasized the importance of macro level safety analysis in transportation planning due to its support of proactive safety planning and management [8]. Researchers have studied relationship between accidents and modal split from other aspects Statistical presentation of fatalities and frequency values such as comparing types of accidents [9], or emphasizing on circumstances of injuries [10]. Other researchers went through registration the traffic flow of motorized vehicles (private vehicles, buses, trucks, power two vehicles) and bicyclists in road segments across streets for an interval of time to figure our pedestrian accidents results, which emphasized that safety issues can change the modal split in favor of vulnerable road users [11]. An article of [12] investigated how travel demand by specific modes of travel in urban areas contributes to crash frequencies at traffic analysis zones (TAZ) level, by developing models that easily could be used for forecasting future crash frequencies, based on future forecasts of trip frequencies. There are several articles discussed forecasting methods; [13] developed the analysis using model to forecast the mortality rates from road accidents using time-series approach, curve estimation, multiple regression analysis, while [14] used The time-series decomposition method. On the other hand, [15] stated that short-term traffic forecasting should not be considered for road accident prediction due to wide range of traffic modelling problems.

Through our research's approach, mode split is considering individuals' differences by applying age intervals unlike the aggregate methods, and connecting road accidents frequencies with modes. Thus, data of passenger's road accidents for a four years' period are gathered. Attributes of mode representation, age, and location are extracted and used. This research establishes a step towards a comprehensive analysis of passengers' behavior in mode choice by studying the accidents they are involved in. Further to that, it could be used to predict future incidents depending on historical accidents data using artificial neural networks, as traffic data were modelled significantly [16].

The structure of the paper is the following: detailed steps of the methodology can be found in Sect. 2. Results of the analysis; significance of data and resulted modal split are presented in Sect. 3. Next is verification of the method in Sect. 4. Finally, conclusion was drawn, and future research trends were identified.

## 2 Methodology

At first, historical accidents data are analyzed for consecutive four years in Hungary. As a total of 103,976 persons were involved in road accidents using different modes their daily traffic extracted from the database of Hungarian Central Statistical Office (KSH). These modes were categorized into 21 classifications whether for passengers' transport or freight. In our case, from the total number of people involved in road accidents, there were 90.7% of them use passenger transport modes, and they have the following role in traffic: front passengers, rear passengers or drivers. This percentage is a very considerable ratio rather any research papers such as [17] which has found that almost half of traffic accidents occurred with passengers in the car. However, each road accident has the following attributes: date of accident, location (cities are used in the research), mode of travel, description of role in the accident (drivers, passengers, pedestrians), and age of persons. Focusing on passengers' travel, they use nine different modes that are represented in road accidents data whether in built up areas or on the open road in cities and towns of Hungary. This means that the classifications are minimized into the following (private cars, motorcycles, bicycles, scooters, buses, trams, rails, suburban rail, trolleybuses). These classifications are categorized into groups later as the analysis needs. Modes of buses, trams, rails, suburban rails, trolleybuses are considered as public transport. To take into consideration individuals' differences, passengers are divided by age into five years' intervals for homogeneity purposes. For each interval, number of passengers involved in road accidents is divided by its corresponding interval population which is conducted from statistical yearbooks of Hungary. Thus, changing these numeric abstract values of in the dataset to a scale without distorting differences in the ranges of values allows to compare more easily these data for different place. Profile of age distribution involved in road accidents using their corresponding modes are explored, and studied through frequency and characteristics of peak age. Statistically; ANOVA (Analysis of variance) is used to test the significance for modes involved in road accidents by both of year and age patterns with 0.05 alpha [18]. Modal split ratios are elaborated for mentioned classifications. Multiple comparisons with real statistical numbers from several articles and publications are used to validate the method. Hypothesis of method validation is carried out for test through P value approach in two ways: overall modal split analysis results by year between passenger cars and public transport, and focused modal split for selected cities of Budapest, Miskolc, Pecs, and Szeged for modes of passenger cars, public transport, and walking/cycling. For forecasting, research uses the Exponential Smoothing (ETS) which is excel function that predicts future values based on existing values that follow a seasonal pattern as a technique of time-series decomposition method. This function calculation depends on triple exponential smoothing. This is an algorithm that applies overall smoothing, trend smoothing, and seasonal smoothing. Models generally fall into two categories, aggregate discrete choice models and disaggregate discrete choice models. The two approaches have similar general concept that passengers will maximize their utility by choosing the mode which offers them the best set of features such as fare, time of travel, comfort. The difference is that aggregate models predict the market share of a mode based on the collective behavior of travelers in the market, whereas disaggregate models examine the mode choice of an individual or household [19]. Thus, once this explained methodology, this research's approach is going towards the discrete models which according to [20] provide rich behavioral predictions as this approach may considered as a contribution to the field that fall in discrete choice models.

# 3 Results

## 3.1 Significance of Data

The P value of the analysis of variance for time period of years and the number of accident that passengers involved in is less than the research (α) level of significant. Likewise, F value is larger than F critical. Then, the conclusion is that time period of years had significant effect on the number of accidents that for different passengers' modes, see Table 1.

**Table 1.** One-way ANOVA: number of accidents versus year

| Source of variation | SS | df | MS | F | P value | F crit |
|---|---|---|---|---|---|---|
| Between groups | 8.63E+08 | 8 | 1.08E+08 | 1742.467 | 1.42E-34 | 2.305313 |
| Within groups | 1670849 | 27 | 61883.28 | | | |
| Total | 8.64E+08 | 35 | | | | |

On the other hand, the P value of the analysis of variance for the different age categories and the number of accident that they involved in is less than the research α level of significant. Likewise, F value is larger than F critical. Then, the conclusion is that the age had significant effect on the number of accidents that people of that age category population are involved in, see Table 2. This is convenient with other researches that found interaction between age group, and car type in dependent variables are statistically significant [21].

**Table 2.** One way ANOVA: number of accident involved passengers versus age category

| Source of variation | SS | df | MS | F | P value | F crit |
|---|---|---|---|---|---|---|
| Between groups | 318188.2 | 8 | 39773.53 | 53.02996 | 2.49E-40 | 1.99939 |
| Within groups | 114753 | 153 | 750.0199 | | | |
| Total | 432941.3 | 161 | | | | |

### 3.2 Age Distribution

The first step in this approach is to find the age distribution of passengers that are involved in road accidents for each classification of modes. The following parts represent statistical values, peaks and bottoms of age distribution for all road accidents involve passengers. It is found that passengers using private cars have the most rate of being involved in road accidents of 2622 accidents yearly per 100,000 capita, followed by bicycles, scooters, and motorcycles as 630, 230, and 215 relatively. Modes of public transportation have much lower rates, which is expected as it is the safest mode universally. Sixth degree polynomial functions represent the relationship between modes and age distribution with different coefficients resulting in good regression values larger than 0.8. For private cars, the distribution of road accidents through age is increasing sharply to a peak point in twenties years old as 280 road accidents per 100,000 capita of this age group, then the distribution goes down at a steady rate, as shown in Fig. 1.

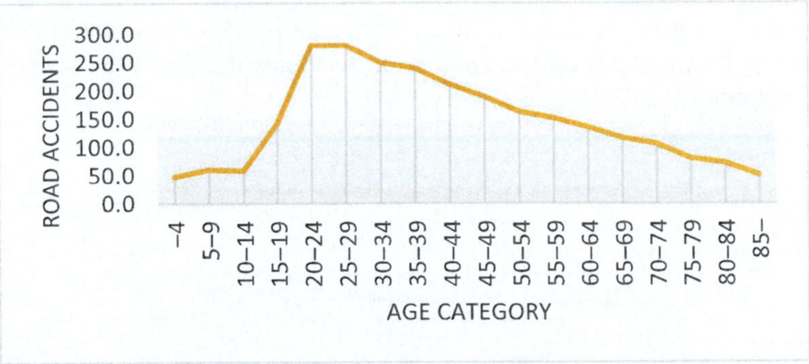


**Fig. 1.** Age distribution for passenger cars in road accidents

As shown in Fig. 2, distributions intersect with each other for different modes. While scooter users are involved more than buses and bicycles users for less than 20 years old, it locates in the middle percentages for age groups between 20 and 50 years old. Reaching 55 years old, all users for the modes of buses, bikes, and scooters are having the same rate of involving in road accidents. Moreover, scooters and buses users are having similar involving ratios for ages more than 60 years old. Bikers are involved in the road accidents more than other modes (except private cars). On the other hand, there is a huge gap in the elderly groups in favor of bicycles, unlike other countries such as Sweden nor Finland

[22]. More specifically, bikers are involved in accidents with a fluctuation rate between 30 and 50 per 100,000 capita in most of group ages. Two peaks are found for groups of 10 to 14 and 50 to 54 years old that reach 50 bikers per 100,000 capita approximately.

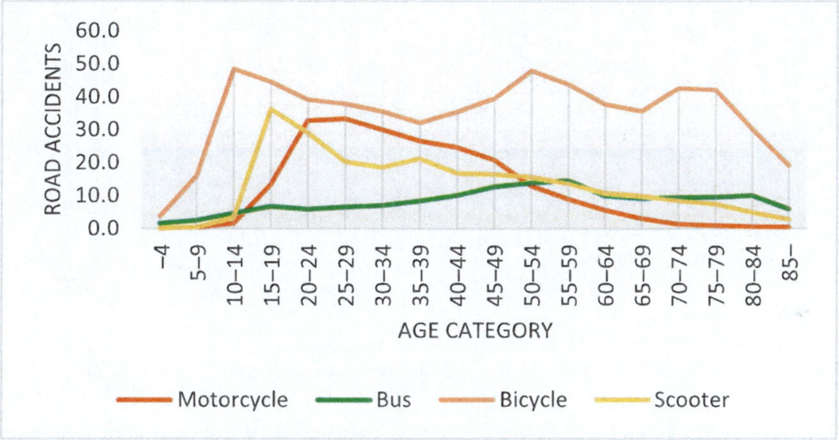


**Fig. 2.** Age distribution for different modes users in road accidents

For motorcycles, the peak is also in twenties years old of 33 road accidents per 100,000 capita of this group age. The concentration of distribution is between 20 and 50 years old, as shown in Fig. 2. On the contrary, scooter users are different from bikers. The distribution reaches a peak point for the group of 15 to 19 years old of a rate as 36 road accidents per 100,000 capita of this group, then the distribution decreases steadily, as shown in Fig. 2. Passengers using buses are involved in road accidents with an increasing rate until 60 years old as a peak point of 15 road accidents per 100,000 for this group, then the rate decreases. This can reflect the dependency of elderly people on public transportation such as buses. The mode classifications of rails, trams, trolleybuses, and suburban rails are categorized as one unit in analysis. The first peak point is 2.3 road accidents per 100,000 capita of the group 45 to 49 years old. The second peak point is 1.7 for the group of 80 to 84 years old.

### 3.3 Modal Split

Subsequent to analyzing the age distribution for each mode, the ratios of passenger modes involved in road accidents for each age category are conducted. The results are shown in Fig. 3. It is found that private cars are the dominant mode with percentage varies from 50% to 90% for different age categories. This high percentage indicates that this mode is the most dangerous in passenger travel. Modes of bicycles and cars are so close in ratios for age category of 10 to 24 years old. Biking ratio is increasing with age as well as bus users on the interpretation of private car users.

To find final modal split ratios for this approach. Percentages of each mode is normalized by population of each age interval. Public transport is categorized as one unit.

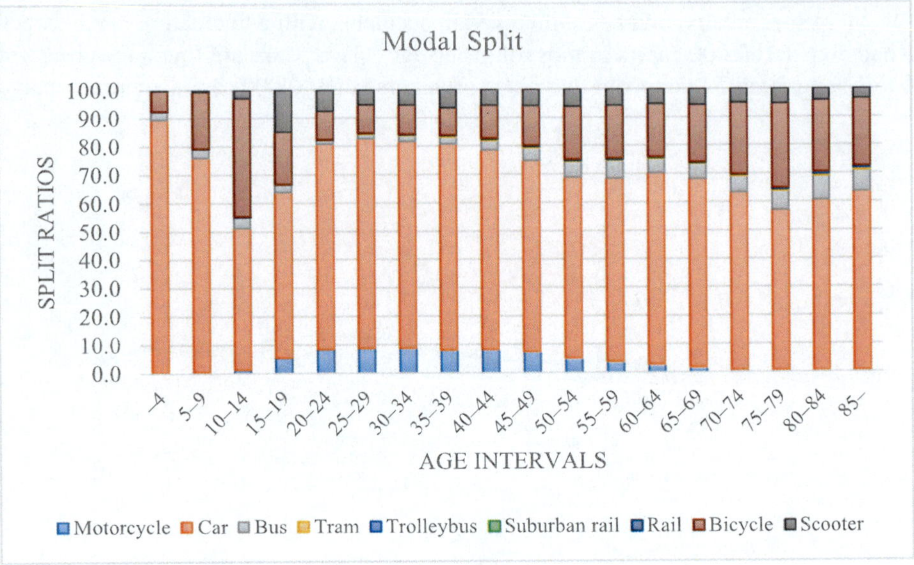


**Fig. 3.** Modal split ratios for passengers transport modes by age intervals

The results are designated in Table 3 for all four years of analysis. Individual years are shown and being compared in verification section.

**Table 3.** Overall modal split ratios for passengers transport modes

| Passengers mode | Modal split ratio |
|---|---|
| Private cars | 67.05% |
| Public transport | 5.99% |
| Cycling | 23.90% |
| Motors | 4.06% |
| **Total** | **100%** |

### 3.4 Forecast of Passenger Cars Split Ratio

Through the data of three consecutive years of modal split ratios for passenger cars, forecast is conducting by following up this period for other four years subjected to time-series decomposition method. Function (ETS) predicts following years' values based on research data. To get the results, function multiplies the seasonality (S) with the trend component (T) which is longer term regular pattern of a time series and thereby differs from the yearly returning (12 month) seasonality pattern. The trend is computed by adding up the least-square regression coefficients after performing a regression analysis

[23]. The green line on the time series plots represents the calculated forecast based on a combination of the yearly regular pattern of seasonality and the long term trend pattern over the study period, extended by four years to draw a prediction of the passenger cars split ratio, see Fig. 4.

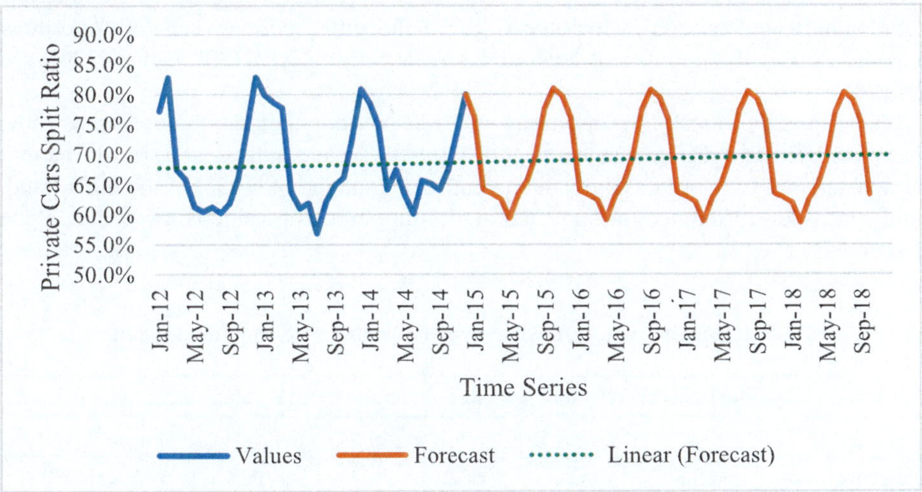


**Fig. 4.** Time series analysis for passenger transport split ratio

# 4 Discussion

## 4.1 Common Remarks

It is found that elderly people prefer to use public transport more than younger. Additionally, younger ages prefer to use non sharing modes such as private cars, scooters, or motorcycles. Bikers are variated through all age categories. The peak of age distribution is moving depending on mode representation; it is 25 for private care, 27 for motorcycles, 12 and 52 for bikers, 17 for scooters, 47 for rails, and 60 years old for buses. Modal split results show that public transport percentages are not that high, which is reflect the safety of these modes not the real representation. Modal split ratios are tested in the next section.

## 4.2 Verification

To find if the results of mode split through this approach are verified or not, values are compared to static numbers from several sources such as official statistical data, articles and publications. In the first way, modal split is compared and tested statistically with overall modal split in Hungary for each corresponding year as proportions of passenger cars and public transport only. In the second way, modal split considers cycling and walking as modes unlike the pervious comparison for selected cities in Hungary.

• Overall Modal Split:

To investigate mode split in Hungary for road networks, data from [24] are used in comparison. They adopt that modal split of inland passenger transport identifies transport by passenger car, by motor coach, bus and trolley bus, and by train [25]. Modal split data from different years are compared with research results with different populations [21–24] and tested through the hypothesis (H0) of that difference between the proportions is equal to 0, with significance level alpha = 0.05 using XLSTAT features [26], see Table 4.

Subsequently to testing proportions with relevant populations, the outcomes show that passenger cars differences between EUROSTAT data and research results cannot be rejected in all four years as p value is higher than alpha of 0.05. On the other hand, public transport differences are rejected in all four years as p value is lower than alpha desperately.

**Table 4.** Significance of research hypothesis through overall modal split

| | EUROSTAT ratios | Research ratios | P value | |
|---|---|---|---|---|
| Year: 2011 | | | | |
| Passenger cars | 68.3 | 68.2 | 0.755 | Accept H0 |
| Public transport | 31.7 | 4.3 | <0.0001 | Reject H0 |
| Year: 2012 | | | | |
| Passenger cars | 67.6 | 67 | 0.059 | Accept H0 |
| Public transport | 32.4 | 4.3 | <0.0001 | Reject H0 |
| Year: 2013 | | | | |
| Passenger cars | 67.6 | 67.4 | 0.53 | Accept H0 |
| Public transport | 32.4 | 4.5 | <0.0001 | Reject H0 |
| Year: 2014 | | | | |
| Passenger cars | 67.5 | 68.1 | 0.052 | Accept H0 |
| Public transport | 32.5 | 4.5 | <0.0001 | Reject H0 |

• Modal split in selected cities considering all modes:

Considering walking and bikes as modes, the modal split ratios are investigated from [5] data to be compared with research results, and tested through the hypothesis (H0) of that difference between the proportions is equal to 0, with significance level alpha equals to 0.05 using XLSTAT features [26]. Subsequently to testing proportions with relevant populations, the outcomes show that walking and cycling differences between EPOMM data and research results cannot be rejected in four out of five cases as p value is higher than alpha of 0.05 in these four cities. On the other hand, passenger cars and public transport differences are rejected in all cases as p value is lower than alpha desperately (Table 5).

**Table 5.** Significance of research hypothesis through all modes

| | EPOMM ratios | Research ratios | P value | |
|---|---|---|---|---|
| Year: 2011, City: Szeged, Population: 161,1631 [27] | | | | |
| Passenger cars | 22.0% | 57.8% | <0.0001 | Reject H0 |
| Public transport | 47.0% | 13.5% | <0.0001 | Reject H0 |
| Walking and cycling | 31.0% | 28.7% | 0.314 | Accept H0 |
| Year: 2014, City: Szeged, Population: 161,921 [27] | | | | |
| Passenger cars | 23.0% | 56.8% | <0.0001 | Reject H0 |
| Public transport | 38.0% | 7.9% | <0.0001 | Reject H0 |
| Walking and cycling | 39.0% | 35.3% | 0.086 | Accept H0 |
| Year: 2011, City: Pecs, Population: 150,027 [28] | | | | |
| Passenger cars | 35.0% | 72.8% | <0.0001 | Reject H0 |
| Public transport | 41.0% | 6.3% | <0.0001 | Reject H0 |
| Walking and cycling | 24.0% | 20.9% | 0.198 | Accept H0 |
| Year: 2014, City: Budapest, Population: 1,744,665 [29] | | | | |
| Passenger cars | 35.0% | 71.3% | <0.0001 | Reject H0 |
| Public transport | 45.0% | 9.8% | <0.0001 | Reject H0 |
| Walking and cycling | 20.0% | 18.9% | 0.042 | Reject H0 |
| Year: 2014, City: Miskolc, Population: 161,265 [30] | | | | |
| passenger cars | 45.0% | 68.6% | <0.0001 | Reject H0 |
| Public transport | 40.0% | 16.5% | <0.0001 | Reject H0 |
| Walking and cycling | 15.0% | 14.9% | 1.000 | Accept H0 |

- Passenger Cars Split Ratio Forecasting:

Comparing the results of yearly forecasting with real data of modal split ratios, it reveals that the differences are acceptable as all of years are less than 1%, which leads to the fact that it is acceptable to do such forecasting depending on statistical data. Table 6 shows the comparison between forecasting and real modal split data extracted from [24].

**Table 6.** Comparison between real modal split ratios and forecasting for passenger cars

| | Eurostat (%) | Forecast (%) | Difference (%) |
|---|---|---|---|
| 2015 | 68.2 | 68.6 | 0.59 |
| 2016 | 69.2 | 69.1 | 0.14 |
| 2017 | 70.0 | 69.7 | 0.43 |
| 2018 | 70.6 | 70.2 | 0.57 |

## 5 Conclusions

This research develops a new approach of finding the modal split through road accidents statistical data. Once finding the modal split between major modes of transport (Private cars, and public transport), this research's approach is acceptable and verified only for passenger cars split ratios and rejected for public transport. As well, forecasting of passenger cars split ratios is accepted and outlines excellent results. On the other hand, the approach is acceptable only for walking and cycling mode, and rejected for passenger cars and public transport if calculating the modal split between all passenger modes (Private cars, public transport, walking and cycling). This leads to the fact that in all cases, public transportation is not applicable in calculating modal split through road accidents statistical data. In general, the approach succeeds partially to measure and forecast modal split ratios of passenger cars, as well as walking and cycling. In the contrary, it fails to measure and forecast public transport slit ratio. Thus, researchers could depend on the accidents data to figure out the modal split ratios in such a year or in forecasting. Future research shall take into consideration rural areas, or small urban context to find the applicability of this approach depending not only on age distribution but other characteristics and attributes of passengers. Freight transport would be an addition.

## References

1. Ungvarai, A.: Modal Split – different approaches to a common term. In: IOP Conference Series: Materials Science and Engineering, vol. 603, p. 042091 (2019). https://doi.org/10.1088/1757-899x/603/4/042091
2. BFCG. Four-Step Travel Demand Model (2020). https://bfcog.us/transportation/gismapping-modeling/regional-travel-demand-model/four-step-travel-demand-model
3. The Future of Interurban Passenger Transport. Report No. 2070-8270, (Organisation for Economic Co-Operation and Development (OECD) (2010)
4. Gaal, G., Horváth, E., Török, Á., Csete, M.: Analysis of Public Transport Performance in Budapest, Hungary. Periodica Polytechnica Social and Management Sciences **23**, 68–72 (2015). https://doi.org/10.3311/ppso.7724
5. EPOMM. The EPOMM modal split tool (2019). http://www.epomm.eu/tems/about_tems.phtml
6. Inter-university Consortium for Political and Social Research ICPSR. Mode Split (2020). https://www.icpsr.umich.edu/CrimeStat/files/CrimeStatChapter.15.pdf

7. Cingel, M., Čelko, J., Drličiak, M.: Analysis in modal split. Transp. Res. Procedia **40**, 178–185 (2019). https://doi.org/10.1016/j.trpro.2019.07.028
8. Abdel-Aty, M., Siddiqui, C., Huang, H., Wang, X.: Integrating trip and roadway characteristics to manage safety in traffic analysis zones. Transp. Res. Record J. Transp. Res. Board **2213**, 20–28 (2011). https://doi.org/10.3141/2213-04
9. Glász, A., Juhász, J.: Car-pedestrian and car-cyclist accidents in Hungary. Transp. Res. Procedia **24**, 474–481 (2017). https://doi.org/10.1016/j.trpro.2017.05.085
10. Duan, A., et al.: A 6-year survey of road traffic accidents in Southwest China: emphasis on traumatic brain injury. J. Saf. Res. **73**, 161–169 (2020). https://doi.org/10.1016/j.jsr.2020.02.010
11. Galanis, A., Botzoris, G., Eliou, N.: Pedestrian road safety in relation to urban road type and traffic flow. Transp. Res. Procedia **24**, 220–227 (2017). https://doi.org/10.1016/j.trpro.2017.05.111
12. Mohammadi, M., Shafabakhsh, G., Naderan, A.: Effects of modal shares on crash frequencies at aggregate level. Accid. Anal. Prev. **120**, 295–303 (2018). https://doi.org/10.1016/j.aap.2018.08.019
13. Jomnonkwao, S., Uttra, S., Ratanavaraha, V.: Forecasting road traffic deaths in Thailand: applications of time-series, curve estimation, multiple linear regression, and path analysis models. Sustainability **12**, 395 (2020). https://doi.org/10.3390/su12010395
14. Timmermans, C., et al.: Analysis of road traffic crashes in the State of Qatar. Int. J. Injury Control Safety Promot. **26**, 242–250 (2019). https://doi.org/10.1080/17457300.2019.1620289
15. Pavlyuk, D.: Feature selection and extraction in spatiotemporal traffic forecasting: a systematic literature review. Eur. Transp. Res. Rev. **11**(1), 1–19 (2019). https://doi.org/10.1186/s12544-019-0345-9
16. Ghanim, M.S., Abu-Lebdeh, G., Ahmed, K.: 2013 5th International Conference on Modeling, Simulation and Applied Optimization (ICMSAO). IEEE 2013 (2013)
17. Hamed, M.M., Al-Eideh, B.M.: An exploratory analysis of traffic accidents and vehicle ownership decisions using a random parameters logit model with heterogeneity in means. Anal. Methods Accident Res. **25** (2020). https://doi.org/10.1016/j.amar.2020.100116
18. Craparo, R.M.: Significance level. In: Encyclopedia of Measurement and Statistics, pp. 889–891 (2007). https://dx.doi.org/10.4135/9781412952644
19. Ashalatha, R., Manju, V.S., Zacharia, A.B.: Mode choice behavior of commuters in Thiruvananthapuram city. J. Transp. Eng. **139**, 494–502 (2013). https://doi.org/10.1061/(asce)te.1943-5436.0000533
20. Garrow, L.A., Koppelman, F.S., Nelson, B.L.: Efficient estimation of nested logit models using choice-based samples. Transp. Traffic Theory 525–544 (2005). https://doi.org/10.1016/b978-008044680-6/50029-5
21. Soliman, A., Alhajyaseen, W., Alfar, R., Alkaabi, I.: Changes in driving behavior across age cohorts in an Arab culture: the case of state of Qatar. Procedia Comput. Sci. **130**, 652–659 (2018). https://doi.org/10.1016/j.procs.2018.04.116
22. Peltola, H., Luoma, J.: Comparison of road safety in Finland and Sweden. Eur. Transp. Res. Rev. **9**(1), 1–13 (2016). https://doi.org/10.1007/s12544-016-0220-x
23. Makridakis, S.: A survey of time series. Int. Stat. Rev. Revue Internationale de Statistique **44**, 29 (1976). https://doi.org/10.2307/1402964
24. EUROSTAT. Modal split of passenger transport (2020). https://ec.europa.eu/eurostat/databrowser/view/tran_hv_psmod/default/table?lang=en
25. European Commission, EC. Mobility and Transport (2020). https://ec.europa.eu/transport/themes/modal-split_de
26. XLSTAT (Addinsoft, 2020)
27. Hungarian Central Statistical Office, H. C. S. Szeged (2020). http://www.ksh.hu/apps/hntr.telepules?p_lang=EN&p_id=33367

28. Hungarian Central Statistical Office, H. C. S. Pécs (2020). http://www.ksh.hu/apps/hntr.telepules?p_lang=EN&p_id=19415
29. Hungarian Central Statistical Office, H. C. S. Budapest (2020). http://www.ksh.hu/apps/hntr.telepules?p_lang=EN&p_id=13578
30. Hungarian Central Statistical Office, H. C. S. Miskolc (2020). http://www.ksh.hu/apps/hntr.telepules?p_lang=EN&p_id=30456
31. Hungarian Central Statistical Office, H. C. S. Statistical Yearbook of Hungary 2012. (KSH, Budapest, 2012)
32. Hungarian Central Statistical Office, H. C. S. Statistical Yearbook of Hungary 2013 (KSH, Budapest, 2013)
33. Hungarian Central Statistical Office, H. C. S. Statistical Yearbook of Hungary 2014 (KSH, Budapest, 2014)
34. Hungarian Central Statistical Office, H. C. S. Statistical Yearbook of Hungary 2011 (KSH, Budapest, 2011)

# Environmental Performance Assessment of Urban Roundabouts

Francesco Acuto[1], Tullio Giuffrè[2], and Anna Granà[1](✉)

[1] Department of Engineering, University of Palermo, Palermo, Italy
{francesco.acuto,anna.grana}@unipa.it
[2] University of Enna Kore, Enna, Italy
tullio.giuffre@unikore.it

**Abstract.** To meet continuous and sustainable improvement on road infrastructures, new technologies and smart devices are encouraging the development of a mobile crowdsensing system for collecting and monitoring data to be used in efficiency assessment. In this view, the paper proposes a methodological approach that employs in an integrated way vehicle trajectory data collected in the field, the Vehicle Specific Power methodology and a microscopic traffic simulation model for estimating emissions at urban roundabouts. For a first characterization of the emissive phenomenon reference was made to an existing two-lane roundabout in Palermo, Italy, where vehicle trajectory data were collected using a smartphone app in a test vehicle. The case study addresses the issue of converting existing roundabouts into turbo roundabouts, and gives an insight into how to assess the impacts of alternative designs from an environmental perspective.

**Keywords:** Roundabout · Turbo roundabout · Pollutant emissions · VSP methodology · Micro-simulation

## 1 Introduction

### 1.1 Background

Smart, resilient and green infrastructures are expressions commonly used to denote enhancing the efficiency of any target infrastructure with the help of the new-generation IoT-based devices and the most modern ICT tools so as to establish new community services. When applied to road infrastructures, the novelty relates to the use of advanced technologies as means to ensure an optimal resource consumption from the project conceptualization to management, and an effective reduction of the health impacts resulting from air pollution, noise, crashes, climate change and territorial degradation during the useful life. It has to be said that there is an increasing discussion on how these notions may relate to each other both at network and single unit level, and may apply to new and existing road infrastructures to meet the decarbonization goal as required by Agenda 2030 ONU and the European Green Deal [1]. In order to foster the emission reduction from the transport sector by 2050, as the EU Commission recommends, emissions from motorized vehicles should be detected and monitored by learning them from vehicle


G. Sierpiński (Ed.): TSTP 2021, LNNS 352, pp. 27–45, 2022.
https://doi.org/10.1007/978-3-030-91156-0_3

trajectory data through new technologies and smart devices that are also encouraging the development of mobile crowdsensing systems [2]. In the transition from the linear to the circular economy model, particular attention should also be given to the tools to be used to assess and to monitor the impact of new infrastructural projects, existing roads or intersections also from an environmental point of view [3].

Among the road intersections, the traffic calming-related effects on vehicle speeds at roundabouts have motivated their widespread use both in urban and rural landscape systems and their integration into environmentally-friendly mobility infrastructures and green spaces [4]. Roundabouts can be traced back to single-lane schemes with one single lane on the entry approaches and circulatory roadway, and to multi-lane counterparts where the entry approaches and circulatory roadway are enough wide to accommodate more vehicles that travel side-by-side [5]. Transition from single-lane to a higher-capacity multi-lane roundabouts has been shown to introduce several safety issues: the increased number of potential conflict points, lane changes within the ring, higher speeds for entering and circulating vehicles, and longer crossing distances for cyclists and pedestrians. Differently from multi-lane roundabouts, the turbo roundabouts better handle safety issues since their spiral geometric design introduces raised lane divisors that guide drivers to select the proper lane before entering to the desired exit [6, 7].

Scientifical and technical literature presents several studies where the performance and safety level of roundabouts and turbo roundabouts have been examined and compared [8, 9]. There are also studies that have assessed the environmental benefits of various types of roundabouts according to the characters of the natural or built context where the roundabouts are operating (or it is expected that they will operate) [10].

Literature refers emission models relied on engine and vehicle activity data to calculate emission rates of different vehicle types and driving conditions that, in turn, are based on wide databases with emission test results; emission rates are also used to provide emission inventories at the national or regional level [11]. However, average emission rates from different vehicles as included in the emission models could predict emissions enough different from those measured in the field [11, 12]. Differently from average speed-based approaches, instantaneous emission models employ vehicle activity parameters to estimate emissions of individual vehicles; see e.g. [13]. It has been found that the interpretation of the emission phenomenon may be rather complex owing to the various parameters as road type and grade, altitude and local temperature, which may affect it [11, 14, 15]. In turn, few studies have been carried out to predict instantaneous emissions by microscopic traffic simulation models especially at intersections and roundabouts. Vasconcelos et al. [16] have assessed the impact on emissions by the curvilinear trajectories through roundabouts. They estimated higher $CO_2$ emissions as a consequence of the conversion of a multi-lane roundabout into a turbo roundabout, pointing out that the turbo roundabout offered advantages for other local pollutants. Although the authors stressed the need to examine more cases, the main novelty of the research was the use of a microsimulation-based approach to support project-level analysis. Fernandes et al. [17] used the Vehicle Specific Power methodology for estimating pollutant emissions from speed profiles through existing multilane-lane and turbo roundabouts, and emissions rates as proposed by VSP mode [13]. Their results showed that, depending on the pollutant, turbo roundabouts produced somewhat higher emissions than multi-lane roundabouts.

### 1.2 The Aim of the Paper

Based on the above, the paper proposes a methodological approach that employs in an integrated way vehicle trajectory data collected in the field, the Vehicle Specific Power (VSP) methodology, and a microscopic traffic simulation model in order to estimate emissions at urban roundabouts. For a first characterization of the emissive phenomenon reference was made to an existing two-lane roundabout in Palermo, Italy, where vehicle trajectory data were collected using a smartphone app in a test vehicle.

Measurements from each trajectory through the existing two-lane roundabout were investigated for parameter calibration in AIMSUN [18]. Model calibration ensured that the simulated speed-travel time profiles could agree properly with the field ones. Specifically, this case study addresses the issue of converting existing roundabouts into turbo roundabouts having comparable size, and gives an insight into how to assess the impacts of an alternative roundabout design from an environmental perspective.

In order to compare the emissions from vehicles moving through the roundabout and the turbo roundabout which was conceptualized with similar size as the existing one, pollutant emissions for each roundabout were estimated based on the VSP modal emission rates and the distribution of time spent in each VSP mode from the speed profiles that were simulated in AIMSUN.

## 2 Data and Methodology

### 2.1 Field Data Collection

A two-lane four-legged roundabout (only roundabout from here on) that operates in the road network of Palermo City, Italy, was the starting point for the study. It was identified among the urban roundabouts of the pilot sample examined by the Authors where pollutant emissions were estimated based on vehicle trajectory data collected by a smartphone app in a test vehicle and the VSP methodology [19]. Entirely consistent with the Italian standards on geometric design of road interchanges and intersections [20], the outer diameter of the existing roundabout is 80 m long, the entry and exit lanes are 4.50 m wide, and the ring roadway is 8.00 m wide within kerbs; the urban limit is 50 km/h.

At the roundabout there were hourly patterns of traffic flows having two distinguishable peaks in the morning and the late afternoon, but no significant peak at noon or in the late evening. Vehicle counting and traffic flow measurements were done during the morning (7:00–8:00 a.m.) and afternoon peak hours (6:00–8:00 p.m.) on weekdays (Thursday to Friday) in July 2020 after the lockdown due to the COVID-19 emergency. It was noted that the morning peak was often reached over a short time period that suddenly dropped to its lowest point, while the afternoon peak was reached over a longer time period than the morning peak. Based on what observed, it was decided to exclude the morning peak hour data compared to data recorded during the afternoon peak hours.

Traffic flow data were both manually and simultaneously counted by two operators, and videotaped on each entry (exit) approach within the system's viewable area; they were recorded by travel direction and classified by manoeuvre so that they can be then identified in the network model of the roundabouts built in AIMSUN. A total entry flow of 3324 vehicles per hour was registered during surveys where about 10% made by heavy vehicles; pedestrian flows were insignificant during the surveys. Since grade is less than 2% this parameter was neglected when the VSP methodology was applied (see next section). Taking into account the available space and the pre-existing constraints of the built environment, a turbo scheme with the inner radius of 15 m was conceptualized based on [5, 21, 22]. Figure 1 shows the layouts of both roundabouts, while Table 1 lists the main geometric features of the turbo roundabout design.

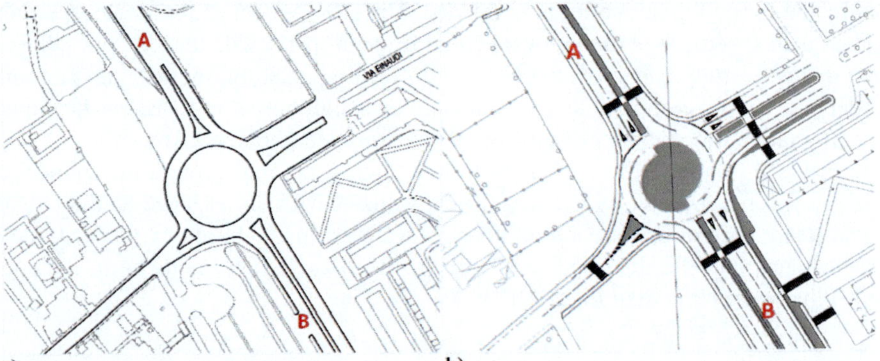


**Fig. 1.** The geometric design of the existing roundabout (a) and the conceptualization of the turbo roundabout layout (b). *Note: A and B only denote the legs interested by the examined through movements (A to B and B to A).*

Data collection also included the recording and analysis of vehicle trajectory data through the roundabout shown in Fig. 1a. As done in a previous research [19], data on location, travel time, distance, grade, second-by-second speed and acceleration were again collected by using the app Speedometer GPS PRO for Android smartphone installed in a Euro IV light passenger diesel vehicle available as the test vehicle. Figure 2 shows the types of trajectories (with no-stop, one-stop and multiple-stops) that were recorded at the sample of roundabouts in Palermo, Italy as referred by [19]. According to [17, 23], vehicles entering a roundabout can experience speed profiles without stop, with one stop at the entry approach or multiple stops as the vehicle joins the queue at the entry leg before making the entry manoeuvre, whose probability of occurrence is closely depending on the intersection traffic congestion level.

In relation to the amount of traffic volume observed during surveys at the roundabout under examination, the test vehicle in most of the GPS runs did not stop completely just before reaching the entry line in order to give priority to circulating vehicles on the ring. Based on what observed in the field at the existing roundabout, the type of speed profile without stop was considered in order to explain the proposed methodology as introduced before. Three GPS runs per driving direction were extracted from the trajectory data that

**Table 1.** The turbo roundabout geometry.

| Cross section elements and roadway edges [m] | | | | | |
|---|---|---|---|---|---|
| Inner radius[a] | 15.0 | Outside roadway width | 5.25 | $R_4 = R_3$ + outside roadway width | 25.925 |
| Inner edge line offset | 0.45 | $Shift_1$ (inside to the middle) | 5.55 | Arc centre bias ($r_1$) | 2.775 |
| Divider inner line offset | 0.20 | $Shift_2$ (middle to the outside) | 5.30 | Start position | 12.225 |
| Divider width | 0.30 | $Bias_1$ ($shift_1/2$) | 2.775 | End position | 17.775 |
| Divider outer line offset | 0.20 | $Bias_2$ ($shift_2/2$) | 2.65 | Arc centre bias ($r_2$) | 2.65 |
| Outer edge line offset | 0.45 | $R_2 = R_1$ + inside roadway width-bias difference | 20.37 | Start position | 17.725 |
| Inside roadway width | 5.50 | $R_3 = R_2$ + divider width | 20.675 | End position | 23.025 |

[a] *inside circulatory roadway (inner edge)*

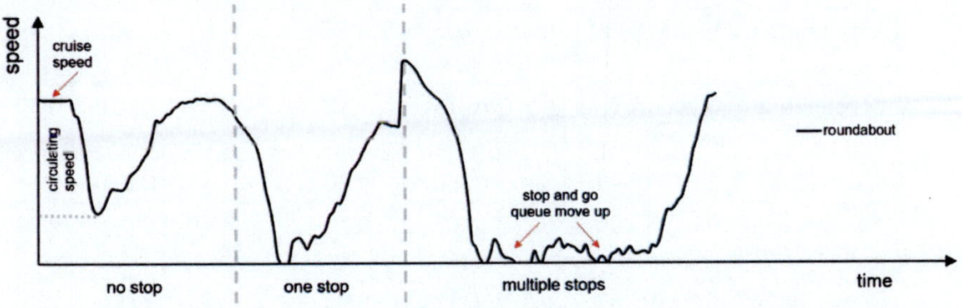


**Fig. 2.** Most common speed profiles observed at roundabouts in Palermo, Italy [19]

were collected by the test vehicle entering the roundabout in Fig. 1a from the left lane for a total of 15 km. Based on second-by-second GPS speed data collected in the field, the acceleration was computed as follows:

$$a_{t_2} = \frac{v_{t_2} - v_{t_1}}{t_2 - t_1} \tag{1}$$

where $a_{t_2}$ is the acceleration at time $t_2$; $v_1$ is the speed at time $t_1$,while $v_2$ is the speed at time $t_2$. According to [24], the acceleration was deemed to be finished when the increase in speed between two successive time points resulted less than 0.1 m/s for next 5 s. The deceleration was also computed based on second-by-second GPS speed data:

$$d_{t_2} = \frac{v_{t_2} - v_{t_1}}{t_2 - t_1} \tag{2}$$

where $d_{t_2}$ is the deceleration at time $t_2$, while $v_1$ and $v_2$ are defined above. Deceleration was calculated from the time at which its values from Eq. 2 resulted greater than or equal to 0.1 $m/s^2$ for following 5 s [24].

It should be noted that the curvilinear paths in the AB and BA directions where the test vehicle experienced the trajectories in the field, were found completely similar (see Fig. 1a). In order to determine if there were statistically significant differences between the two driving directions a two-tailed t-test was done. The results in Table 2 show that no significant difference existed between the AB and BA directions and the null hypothesis could not be rejected.

**Table 2.** The two-tailed t-test for the relevant parameters detected in the field.

| Profile | Parameter | $\mu_{AB}$[a] (s.e.) | $\mu_{BA}$[a] (s.e.) | t-value[b] ($\alpha$ = 0.05) | n[c] | $t_{crit}$ | p-t test |
|---|---|---|---|---|---|---|---|
| 1 | Speed (m/s) | 9.211 (0.60) | 7.457 (0.80) | 1.73 | 56 | 2.008 | 0.10 |
| | Acceleration ($m/s^2$) | 0.473 (0.06) | 0.373 (0.05) | 1.16 | 61 | 2.005 | 0.26 |
| | Deceleration ($m/s^2$) | 0.485 (0.14) | 0.453 (0.10) | 0.19 | 53 | 2.013 | 0.86 |
| 2 | Speed (m/s) | 8.613 (0.70) | 9.02 (0.85) | −0.36 | 46 | 2.015 | 0.72 |
| | Acceleration ($m/s^2$) | 0.448 (0.08) | 0.463 (0.06) | −0.15 | 60 | 2.002 | 0.88 |
| | Deceleration ($m/s^2$) | 0.750 (0.18) | 0.502 (0.13) | 1.12 | 47 | 2.030 | 0.30 |
| 3 | Speed (m/s) | 8.283 (0.48) | 8.493 (0.90) | −0.20 | 35 | 2.030 | 0.84 |
| | Acceleration ($m/s^2$) | 0.510 (0.06) | 0.495 (0.08) | 0.14 | 49 | 2.009 | 0.89 |
| | Deceleration ($m/s^2$) | 0.40 (0.06) | 0.577 (0.18) | −0.92 | 34 | 2.032 | 0.40 |

[a]$\mu_{AB}$ and $\mu_{BA}$ stand for the mean values of the samples of the relevant parameters in the two driving directions, while s.e. stands for standard error standard error of mean; [b]t-value is the result of the two tailed t-test to compare the equality of the means $\mu_{AB}$ and $\mu_{BA}$ of the samples of two populations at the confidence level $\alpha$; [c]n stands for degree-of-freedom

Table 3 shows a summary of the mean values of the relevant parameters by manoeuvre (i.e. speed, the acceleration and deceleration, the 85$^{th}$ and 95$^{th}$ percentile values of acceleration and deceleration) calculated from the corresponding parameter values extracted from every speed profile experienced by the test vehicle in each run and driving direction (see Fig. 1a).

## 2.2 Emission Estimation Using the VSP Methodology

It is known that the Vehicle Specific Power (VSP) model allows the estimation of the power required to the engine while a vehicle is driving and it describes the dependence

**Table 3.** Summary of the mean values of the vehicle activity parameters by driving direction.

| Parameter | Direction | |
|---|---|---|
| | AB | BA |
| Max. speed (m/s) | 15.01 | 14.99 |
| Max. acceleration (m/s$^2$) | 1.46 | 1.60 |
| Max. deceleration (m/s$^2$) | 2.48 | 3.40 |
| 85$^{th}$ percentile acceleration (m/s$^2$) | 0.83 | 0.80 |
| 95$^{th}$ percentile acceleration (m/s$^2$) | 1.10 | 1.09 |
| 85$^{th}$ percentile deceleration (m/s$^2$) | 1.13 | 0.94 |
| 95$^{th}$ percentile deceleration (m/s$^2$) | 1.58 | 1.89 |

of vehicle emissions on the power. According to [13], the VSP defines the instantaneous tractive power per unit vehicle mass and it is used to overcome the aerodynamic resistance force and rolling resistance force, and to increase both the kinetic energy and the potential energy:

$$Vehicle\ Specific\ Power = \frac{\frac{d(E_k+E_p)}{dt} + F_{rolling} \cdot v + F_{\text{aerodynamic}} \cdot v}{m} \quad (3)$$

where the *VSP* is the power per unit mass; $E_k$ and $E_p$ are the kinetic energy and the potential energy, respectively; $F_{rolling}$ and $F_{aerodynamic}$ represent the rolling resistance force and the aerodynamic resistance force, respectively; $v$ is the instantaneous speed; $m$ is the vehicle mass. In the simplified form the VSP equation for a typical light passenger vehicle is based on the road grade, vehicle's speed and acceleration [25, 26]:

$$VSP = v \cdot [1.1 \cdot a + 9.81 \cdot sin(arctan(grade)) + 0.132] + 0.000302 \cdot v^3 \quad (4)$$

where *VSP* is in kilowatts per metric ton; $v$ is the instantaneous speed (m/s), *grade* is the road grade (decimal fraction), and $a$ is the acceleration (m/s$^2$). Note that 14 VSP bins were identified, each representing a power requirements interval; thus, every second of driving can be associated with a VSP bin and the corresponding power requirements interval. Based on the 14 modes of engine regime an instantaneous emission factor by mode can be used to estimate pollutant emissions by vehicle type. By way of examples Table 4 shows the emission rates by VSP mode for light passenger diesel vehicles that were used to calculate emissions for the case study in the following section; the average values for $CO_2$, CO, $NO_x$, and HC emission rates by VSP mode for light passenger diesel vehicles, light passenger gasoline vehicles and light commercial diesel vehicles can be also found in [17, 26].

The VSP values were calculated by Eq. (4) for each driving direction travelled in each speed profile experienced in the field; they were, therefore, used as a joint measure to provide an overall view of the vehicle trajectory data. Figure 3 shows the percentages of the time spent in each VSP mode for the speed profiles experienced by the test vehicle through the existing roundabout. Despite the variability found in the three cases here

**Table 4.** VSP mode, power requirements and mean emission rates by VSP mode for LPDV[a].

| VSP mode | Kw/ton[b] | Emission rates (g/s) | | | |
|---|---|---|---|---|---|
| | | $CO_2$ | CO | $NO_x$, | HC |
| 1 | VSP < −2 | 0.21 | 0.00003 | 0.0013 | 0.00014 |
| 2 | −2 ≤ VSP < 0 | 0.61 | 0.00007 | 0.0026 | 0.00011 |
| 3 | 0 ≤ VSP < 1 | 0.73 | 0.00014 | 0.0034 | 0.00011 |
| 4 | 1 ≤ VSP < 4 | 1.50 | 0.00025 | 0.0061 | 0.00017 |
| 5 | 4 ≤ VSP < 7 | 2.34 | 0.00029 | 0.0094 | 0.00020 |
| 6 | 7 ≤ VSP < 10 | 3.29 | 0.00069 | 0.0125 | 0.00023 |
| 7 | 10 ≤ VSP < 13 | 4.20 | 0.00058 | 0.0155 | 0.00024 |
| 8 | 13 ≤ VSP < 16 | 4.94 | 0.00064 | 0.0178 | 0.00023 |
| 9 | 16 ≤ VSP < 19 | 5.57 | 0.00061 | 0.0213 | 0.00024 |
| 10 | 19 ≤ VSP < 23 | 6.26 | 0.00101 | 0.0325 | 0.00028 |
| 11 | 23 ≤ VSP < 28 | 7.40 | 0.00115 | 0.0558 | 0.00037 |
| 12 | 28 ≤ VSP < 33 | 8.39 | 0.00096 | 0.0743 | 0.00042 |
| 13 | 33 ≤ VSP < 39 | 9.41 | 0.00077 | 0.1042 | 0.00040 |
| 14 | VSP ≥ 39 | 10.48 | 0.00073 | 0.1459 | 0.00042 |

*[a]LPDV stands for Light Passenger Diesel Vehicles; [b]as calculated by Eq. (4)*

examined, the time percentages were on the whole consistent with each other by direction both in the VSP modes 1 to 2 (deceleration) and 4 to 5 (acceleration). It should be noted that the results concern the speed-time profiles corresponding to the distance of 250 m driven by the test vehicle through the existing roundabout; this choice has been made both to obtain speed profiles as coherent as possible with each other, and to extract the contribution of the roundabout to the emission phenomenon (see next section).

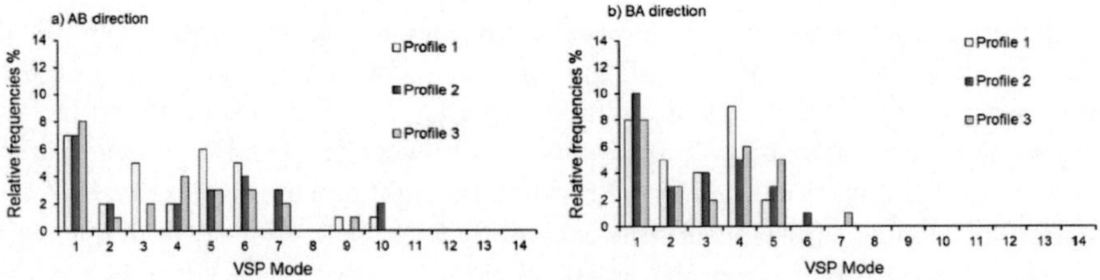


**Fig. 3.** Relative frequencies of time spent in VSP modes

A two-tailed *t-test* (5% significance level) was also performed between the VSP distributions in the AB and BA directions travelled by the test vehicle in each speed profile; since the *p-value* resulted greater than 0.05, there was not enough evidence to reject the null hypothesis that the VSP distributions in the AB and BA directions came from the same population.

It is well known that instantaneous emission models relate emission rates to second-by-second vehicle dynamic data [11]. The instantaneous emission rates were estimated from the distribution of time spent in each VSP mode obtained by the speed profiles under examination. In order to estimate the emissions for each speed profile experienced by the test vehicle through the existing roundabout, the instantaneous emission rates were calculated using the VSP values from Eq. (4). According to [17], the emission by pollutant source and each profile has been obtained by summing all the emission factors (g/s) assigned to the $i^{th}$ second of the speed profile based on the second-by-second VSP mode (see Table 4). Similarly to [19], the total emissions associated with the three profiles (i.e., i = 1, 2 or 3) were then obtained as the average of the emissions per pollutant source and speed profile.

By way of example, Fig. 4 shows the comparison between observed acceleration and deceleration values experienced by the test vehicle through the roundabout (AB movement, profile 1) with $NO_X$ second-by-second grams from Eq. 4 and Table 4. Since the kinematic parameters as accelerations and decelerations may affect emissions generated from vehicle activity, one can observe that, except for very low speeds (i.e., less than 4 km/h) and accelerations under 0.15 m/s$^2$, higher average speeds and the high average accelerations have significant influence on pollutant emissions [27].

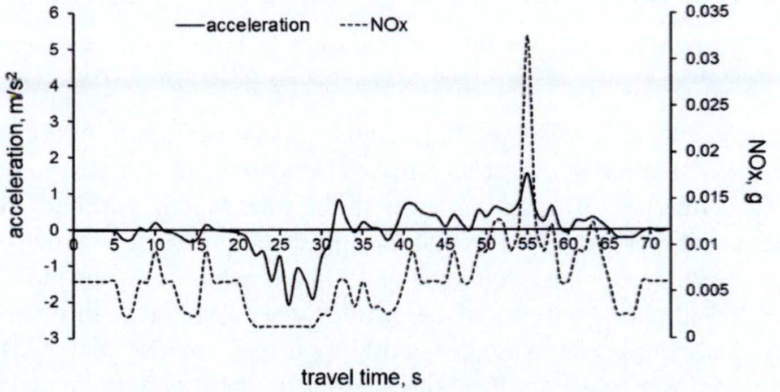


**Fig. 4.** The AB movement-level acceleration and emissions estimates of NOx through the roundabout (profile 1).

## 2.3 AIMSUN Modelling

The AIMSUN software was used to simulate the traffic flows in the same temporal intervals as they were surveyed in the field (see Sect. 2.1); the warm-up time was also set to equalize traffic flows deployed through the O/D matrix. It should be noted that the users can use the circulation rules of the roundabouts within the software editor when the circulatory roadway and entry/exit widths should be modelled as the users want. Figure 5 shows the sketches of the network models of the roundabout and turbo roundabout built in AIMSUN; they were used to insert the input traffic from O/D matrices. The O/D matrices were differentiated in order to include the vehicle percentages as collected in the field; the detectors supporting generation of the second-by-second speed profiles were localized along the vehicle paths through both roundabouts.

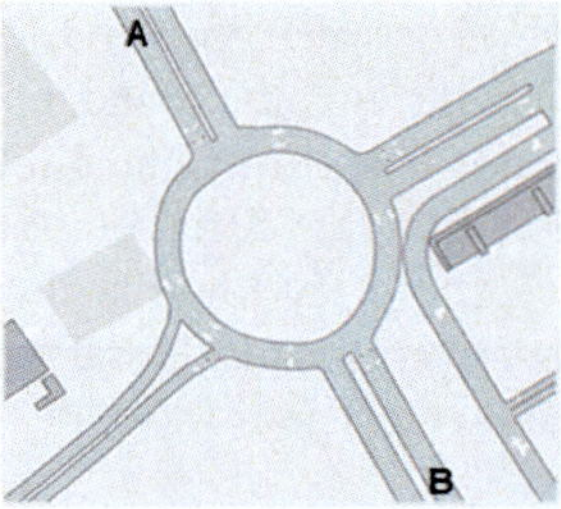


a)

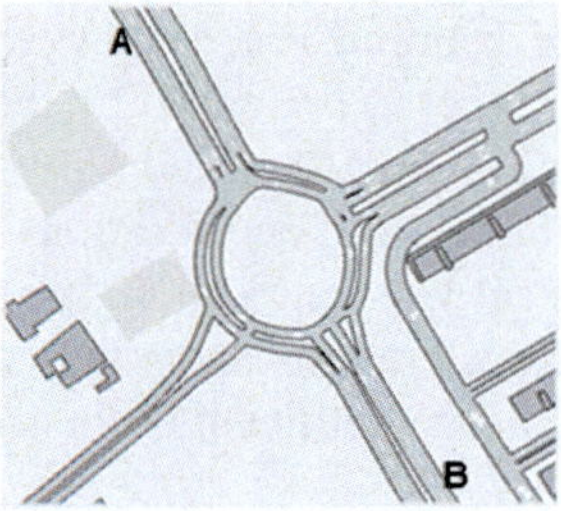


b)

**Fig. 5.** The sketches of the two-lane roundabout (a) and the turbo roundabout (b) in AIMSUN.

Consistently with what observed in the field, the speed limit of 50 km/h was set; it represented the maximum speed at which the roundabout is travelled under free flow conditions. Once the microsimulation ran with the time scan per second, the trajectory data were extracted and processed so that the simulated profiles could be generated.

Before proceeding with the calculation of pollutant emissions using the VSP methodology, the observed and simulated speed profiles were compared. In order to improve the accuracy of the emission estimates, calibration was done so that AIMSUN could return speed profiles as close to those obtained in the field and the model was able to best replicate them. According to an initial sensitivity analysis, the following vehicle attribute parameters of AIMSUN were selected: the *maximum acceleration* (MA), m/s$^2$, or the highest acceleration that a vehicle can achieve under any driving circumstances in the network; the *normal deceleration* (ND), m/s$^2$, or the maximum deceleration of a vehicle under normal conditions. Note that the default values of AIMSUN parameters are equal to 3.00 m/s$^2$ and 4 m/s$^2$ for the maximum acceleration and the normal deceleration, respectively. Based on what explained by Anya et al. [28] for urban arterial corridors, the vehicle attribute parameters of AIMSUN were supplied with the values of the 95$^{th}$ (or 85$^{th}$) percentiles of the accelerations and decelerations extracted from the distributions measured in the field. The following combinations have been examined: a) the 95$^{th}$ (or 85$^{th}$) percentile values of accelerations and decelerations extracted from every field-observed vehicle trajectory; b) the 95$^{th}$ (or 85$^{th}$) percentile values of accelerations and decelerations extracted from all the field-observed vehicle trajectories travelled across the roundabout per driving direction; c) the 95$^{th}$ (or 85$^{th}$) percentile values of accelerations and decelerations extracted from all the field-observed vehicle trajectories travelled across the existing roundabout regardless of the driving direction. The combinations of the AIMSUN parameters under calibration with the mean values of the 95$^{th}$ or 85$^{th}$ percentiles of accelerations and decelerations extracted from all the field-observed trajectories experienced by the test vehicle regardless of the driving direction best fitted the field data and returned VSP distributions more realistic than the others. Under these combinations the model also resulted effective in improving the emission estimates. In order to determine how similar observed and simulated speed profiles were, the GEH index was used as a measure of goodness-of-fit of the simulated data to those observed in the field [29]. Based on current interpretation, each $GEH_i$ was also verified for each pair of simulated and observed speed values concerning every speed profile. Since the deviation of the simulated values with respect to the observed values resulted smaller than 5 in more than the 90% of the cases, the model was accepted. For the purposes of the analysis here presented, the root mean squared normalized errors less than 15 percent were also considered as acceptable.

By way of example, Fig. 6 shows the comparison between the GPS data detected in the field by using a smartphone installed in the vehicle and the speed-time profiles simulated in AIMSUN. Specifically, the figure shows the speed-time profiles simulated under the default parameters and calibration with the mean values of the 85$^{th}$ or 95$^{th}$ percentile accelerations and decelerations extracted from the field-observed GPS trajectories regardless of the driving direction. It should be noted that the vertical dashed lines in Fig. 6 identify the influence area of 250 m (i.e. about three times the diameter) that was used to extract the specific contribution of the roundabout on the emission phenomenon; it includes the deceleration distance of a vehicle travelling from the cruise speed as it approaches and then enters the roundabout, and the acceleration distance as it exits the roundabout up to the section where the cruise speed is about to be reached. For the three profiles examined in this study, lower speed variations resulted under calibration with the 85$^{th}$ percentile values of the relevant parameters than the 95$^{th}$ percentile values.

Starting from the speed profiles simulated in the two driving directions through the existing roundabout in Fig. 1a, the VSP values were calculated by Eq. (4); they were used as a joint measure to provide an overall view of the vehicle trajectory data (i.e. the second-by-second speed and acceleration) so as to compare the results of microsimulation with the observed ones at the existing roundabout. Table 5 shows the results of the two-tailed *t-test* for the observed vs simulated VSP values for the roundabout in Fig. 1a; they were deemed acceptable in view of the purposes of the analysis here presented where only three profiles in both driving directions were examined. It follows that there is no significant difference in the two distributions of the observed and simulated VSP values that have been compared from time to time.

Based on calibration above, analogous speed profiles were simulated for the turbo roundabout in order to compare performances of both roundabouts from an environmental point of view. Figure 7 shows, by way of example, the simulated speed-time profiles under the 95$^{th}$ percentile values of the relevant parameters for the roundabout vs the turbo roundabout. Again, the vertical dashed lines in Fig. 7 identify the influence area of 250 m, while the speed profile identified by the black line represents the speed profile

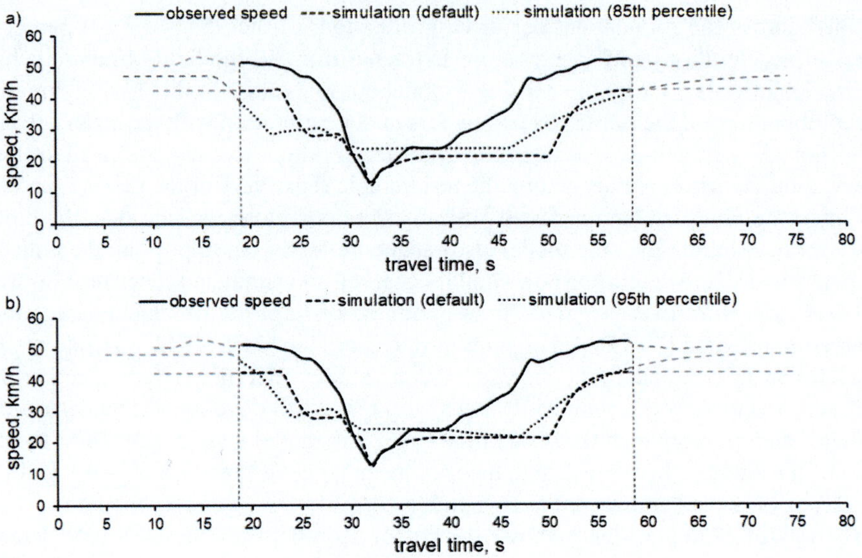


**Fig. 6.** Observed vs simulated speed profiles under (a) the 85th percentile values and (b) the 95th percentile values of the relevant parameters at the roundabout (profile 1, movement A-B)

experienced by the test vehicle through the roundabout in the field. It must be noted that the observed speeds in the figure introduced above concern the existing roundabout since at this stage of analysis the turbo roundabout has been conceptualized only; the observed speeds in Fig. 7 were introduced just as term of comparison for the simulation output of the existing roundabout.

Figure 8 depicts, in turn, an example of scattergram analysis that was done to compare the observed versus simulated speeds under calibration with the mean values of the 95th percentiles of the relevant parameters. The regression lines of the observed versus simulated flows at the detection stations of the roundabout model is plotted along with the 95% prediction interval. Given that the $R^2$ value is 0.936 for the roundabout and only a few points lay out of the confidence band of the prediction intervals, the model output could be considered as close as possible to the reality. Note that the observed speeds in the regression plot correspond to a GPS run made at the two-lane roundabout (i.e. the profile 2 in the BA direction); similar results were obtained also in the other cases.

Figure 9 shows, by way of example, the distributions of time spent in VSP modes under parameters calibrated with the mean values of the 95th percentiles of the accelerations and decelerations extracted from all the vehicle trajectories travelled across the roundabouts; they were obtained by averaging the frequency of the time spent in each VSP mode for the three reference profiles here examined.

Based on [28] the distributions of the 85th percentile accelerations and decelerations could be more representative of the values of accelerations and decelerations observed in the field on signalized arterial corridors; however, sampling from them may constrain the maximum acceleration or deceleration values that can be achievable in AIMSUN and the simulated emissions could result less realistic. On roundabouts, the vehicles

**Table 5.** Two-tailed *t-test* for the observed vs simulated VSP distributions for the roundabout.

| VSP | Direction | $\mu_{AB}$[a] (s.e.) | $\mu_{AB}$[a] (s.e.) | $t_{(\alpha=0.05)}$[b] | n[c] | $t_{crit}$ | p-t test |
|---|---|---|---|---|---|---|---|
| Obs. Vs default | AB | 2.43 (1.26) | 1.67 (1.30) | 0.421 | 76 | 1.992 | 0.675 |
| Obs. Vs default | BA | 2.21 (1.26) | 2.19 (1.538) | 0.112 | 72 | 1.993 | 0.99 |
| Obs. Vs 85$^{th}$ percentile | AB | 2.36 (1.26) | 1.03 (0.77) | 0.90 | 63 | 1.998 | 0.40 |
| Obs. Vs 85$^{th}$ percentile | BA | 2.27 (0.991) | 1.40 (0.524) | 0.777 | 74 | 1.992 | 0.44 |
| Obs. Vs 95$^{th}$ percentile | AB | 2.02 (1.28) | 0.64 (0.90) | 0.828 | 72 | 1.993 | 0.41 |
| Obs. Vs 95$^{th}$ percentile | BA | 2.22 (1.262) | 0.82 (0.969) | 0.877 | 76 | 1.993 | 0.383 |
| *Roundabout (profile 2)* | | | | | | | |
| Obs. Vs default | AB | 2.18 (1.40) | 1.67 (1.30) | 0.262 | 75 | 1.992 | 0.794 |
| Obs. Vs default | BA | 1.08 (1.14) | 2.27 (1.84) | –0.548 | 64 | 1.997 | 0.585 |
| Obs. Vs 85$^{th}$ percentile | AB | 2.26 (1.39) | 0.77 (0.78) | 0.929 | 63 | 1.998 | 0.356 |
| Obs. Vs 85$^{th}$ percentile | BA | 2.04 (1.02) | 1.15 (0.54) | 0.768 | 71 | 1.993 | 0.444 |
| Obs. Vs 95$^{th}$ percentile | AB | 1.94 (1.53) | 0.89 (1.11) | 0.555 | 72 | 1.996 | 0.580 |
| Obs. Vs 95$^{th}$ percentile | BA | 1.59 (1.21) | 0.95 (1.02) | 0.404 | 74 | 1.992 | 0.687 |
| *Roundabout (profile 3)* | | | | | | | |
| Obs. Vs default | AB | 2.15 (1.40) | 1.66 (1.30) | 0.289 | 76 | 1.993 | 0.773 |
| Obs. Vs default | BA | 1.88 (1.29) | 2.27 (1.82) | –0.171 | 68 | 1.995 | 0.865 |
| Obs. Vs 85$^{th}$ percentile | AB | 2.55 (0.98) | 1.45 (0.53) | 0.982 | 65 | 1.997 | 0.329 |
| Obs. Vs 85$^{th}$ percentile | BA | 2.35 (1.12) | 1.05 (0.63) | 1.001 | 72 | 1.993 | 0.320 |
| Obs. Vs 95$^{th}$ percentile | AB | 2.03 (0.97) | 1.28 (0.629) | 0.649 | 70 | 1.994 | 0.518 |
| Obs. Vs 95$^{th}$ percentile | BA | 2.11 (1.30) | 1.12 (1.18) | 0.563 | 75 | 1.992 | 0.575 |

[a] $\mu_{AB}$ and $\mu_{BA}$ stand for the mean values of the samples of the observed and simulated values of VSP (Kw/ton) in the two driving directions, while s.e. stands for standard error of mean; [b] t-value is the result of the two tailed t-test to compare the equality of the means $\mu_{AB}$ and $\mu_{BA}$ of the samples of two populations at the confidence level $\alpha$; [c] n stands for degree-of-freedom

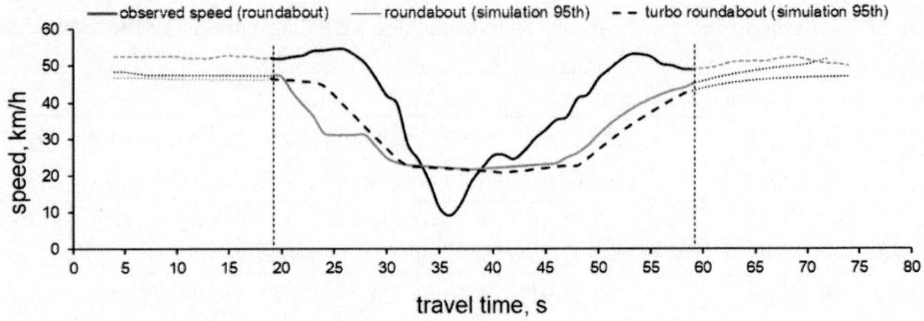


**Fig. 7.** Roundabout vs turbo roundabout simulated speed-time profiles under the 95th percentile values of the relevant parameters (profile 2, movement A-B). *Note: since the turbo roundabout was conceptualizes at this stage of analysis, the observed speeds concern the roundabout only; they were used just as term of comparison for the simulation output of the existing roundabout.*

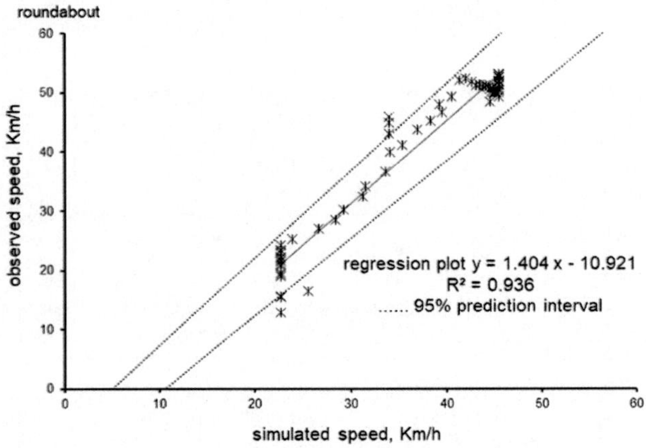


**Fig. 8.** Regression plot to compare observed vs simulated speeds under calibration with 95th percentile values of the relevant parameters (profile 2, movement BA).

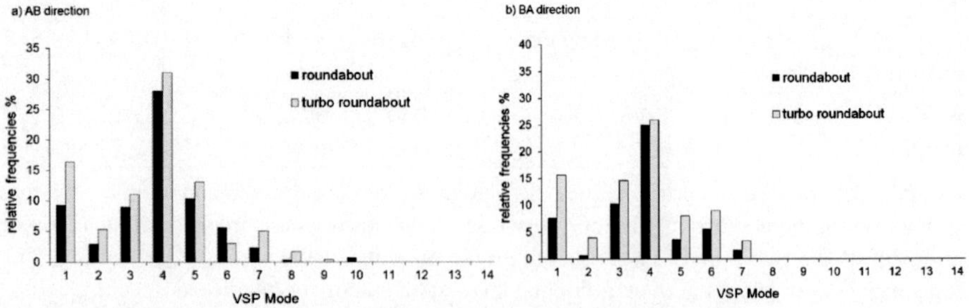


**Fig. 9.** VSP distributions (relative frequency %) from simulated data under calibrated parameters on the roundabout and the turbo roundabout

took longer time to travel the circulatory roadway without significant changes in speed. Calibrating the model with 95$^{th}$ percentile values of the relevant parameters observed in the field at the existing roundabout, in turn, tended to be more effective in producing VSP distributions enough consistent between the two-lane roundabout and the turbo roundabout.

## 2.4 Results

The speed profiles experienced in the field through the existing roundabout and those returned by AIMSUN for both roundabouts were the starting point to estimate emissions. Based on the time spent by the particular vehicle here considered in every VSP mode and the second-by-second emission rates in Table 4, emissions for each source pollutant and speed profile were calculated as introduced in Sect. 2.2. An influence area of 250 m was considered in order to have consistency among the runs and to extract the impact of the roundabouts to the emissive phenomenon. Total emissions were then obtained as average of emission for source pollutant and speed profile. Figure 10 shows the comparison among the emission estimates for the different pollutants for the roundabout and the turbo roundabout. Note that the observed emissions in Fig. 10 refer to the values calculated employing Eq. (4) and the emission rates by VSP mode in Table 4 starting from the speed profiles travelled by the test vehicle across the existing roundabout; they were used as term of comparison for the corresponding simulation output since at this stage of analysis the turbo roundabout was conceptualizes only.

The results confirmed the versatility of AIMSUN in estimating emissions based on instantaneous speed-time profiles and improvements in its predictive capacity when calibration with the 95$^{th}$ percentiles of the relevant parameters is used; in this case, indeed, one can observe that the simulated emissions were closer to those detected in the field and the differences between the simulated values of $CO_2$ and CO emissions between the two roundabouts tended to be lower than other simulations.

It should be noted that results may have been affected by the assumptions on which this study is based concerning the particular test vehicle used in the field (i.e. a light passenger diesel vehicle), the design options under examination and the calibration process performed to identify and to adjust parameters that enable the model to better reproduce the observed emissions. Besides believing that the study should be extended to a larger sample of observations, it is also important to stress the chance to examine in the near future the role of a calibration process that includes the "global parameters" or further "local parameters" of AIMSUN for their possible effects on the simulated vehicle activity, in combination with the "vehicle attributes" that were here fine-tuned only based on the relevant parameter values extracted from the distributions surveyed in the field.

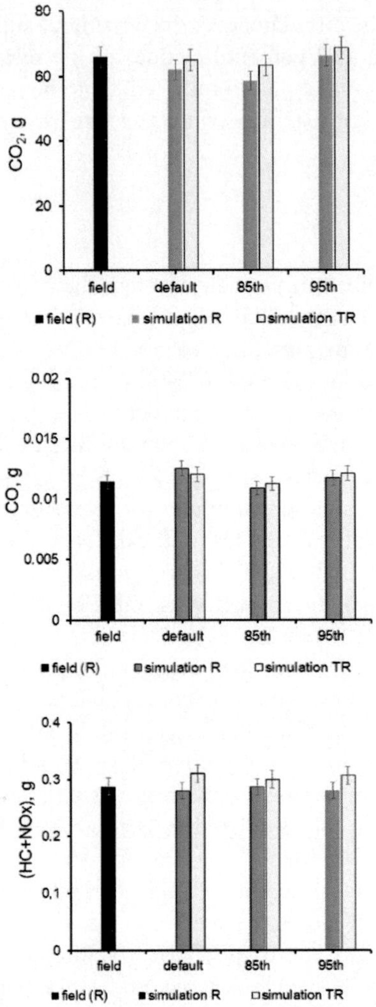


**Fig. 10.** Total emission [g] for the roundabout and the turbo roundabout. *Note: R stands for roundabout, while TR stands for turbo roundabout; the observed speeds concern the roundabout and they were used just as term of comparison for the simulation output of the existing roundabout since the turbo roundabout was conceptualizes only at this stage of analysis.*

## 3 Discussion and Conclusions

The paper addresses the feasibility of converting an existing roundabout into a turbo roundabout and gives insights into how to assess the impacts of alternative roundabout designs from an environmental perspective. This study employed the Vehicle Specific Power (VSP) methodology which uses speed trajectories to estimate the second-by-second emissions generated from vehicles and the AIMSUN software to simulate speed-travel time profiles travelled across the roundabouts. Reference was made to a case study of an existing two-lane roundabout in Palermo, Italy, where vehicle trajectory data along with traffic volumes were collected. Model calibration was also made based on the comparison between the individual GPS trajectories on-field collected, and second-by-second speed profiles derived from AIMSUN.

In order to quantify the emission impacts and compare the emissions from vehicles moving through the two-lane roundabout and the turbo roundabout, pollutant emissions were estimated from the VSP modal emission rates and the distribution of time spent in each VSP mode from the speed profiles that were simulated in AIMSUN.

According to literature [16, 17] the results showed that that travel times spent in acceleration were somewhat higher through the turbo roundabout than the two-lane roundabout; thus, the implementation of a turbo roundabout could offer no significant advantage if the major concern is the reduction in emissions. Although in this case study the travel times spent in acceleration resulted on average slightly higher through the turbo roundabout than the two-lane roundabout, the conversion of a two-lane roundabout to a turbo roundabout resulted in a fully comparable amount of emissions. This primarily is due to the low volume conditions observed in the field where the two-lane roundabout is still the more appropriate solution. Thus, when it is necessary to design a two-lane roundabout also as an alternative to a single-lane roundabout, the turbo roundabout option is probably the best solution from a safety point of view only, unless a maximum output of capacity is also expected and a multi-lane roundabout can be still the preferred option. However, further potential trade-offs should be also considered owing to the variety of real-world situations that could lead to review the choice of installing roundabouts versus other types of intersections [30].

At last, the main finding provided from this case study is referred to the feasibility of the methodological approach that employs in an integrated way vehicle trajectory data collected in the field using a smartphone app, the VSP methodology, and a microscopic traffic simulation model in order to estimate emissions at urban roundabouts. It highlights the positive potential of a novel attitude in the conceptualization and performance evaluation of road units in order to align urban infrastructural projects with the shared long-term ambitions about a low-emission urban mobility. It is also feasible the use of smartphone for vehicle activity data collection, which will enable to reduce the cost of on-field surveys and to increase significantly the data volume that can be collected and shared through the interested user's community.

## References


1. Haas, T., Hendrik, S.: Decarbonizing transport in the European union: emission performance standards and the perspectives for a European green deal. Sustainability **12**(20), 8381 (2020).https://doi.org/10.3390/su12208381
2. Capponi, A., Fiandrino, C., Kantarci, B., Foschini, L., Kliazovich, D., Bouvry, P.: A survey on mobile crowdsensing systems: challenges, solutions, and opportunities. IEEE Commun. Surv. Tutor. **21**(3), 2419–2465 (2019)
3. Girard, L.F., Nocca, F.: Moving towards the circular economy/city model: which tools for operationalizing this model? Sustainability **11**(22), 6253 (2019). https://doi.org/10.3390/su11226253
4. Dall'Ara, E., Maino, E., Gatta, G., Torreggiani, D., Tassinari, P.: Green mobility infrastructures. a landscape approach for roundabouts' gardens applied to an Italian case study. Urban Forestry Urban Greening **37**, 109–125 (2019)

5. National Academies of Sciences, Engineering, and Medicine, Roundabouts: An Informational Guide (2010). https://doi.org/10.17226/22914
6. Fortuijn, L.G.H.: Turbo roundabouts: design principles and safety performance. Transp. Res. Rec. **2096**(1), 16–24 (2009)
7. Tollazzi, T.: Alternative Types of Roundabouts: An Informational Guide. Springer, New York (2015). https://doi.org/10.1007/978-3-319-09084-9
8. Macioszek, E.: The road safety at turbo roundabouts in Poland. Arch. Transp. **33**(1), 57–67 (2015)
9. Corriere, F., Rizzo, G., Guerrieri, M.: Estimation of air pollutant emissions in "turbo" and in conventional roundabouts. Appl. Mech. Mater. **394**, 597–604 (2013)
10. Guerrieri, M., Corriere, F., Lo Casto, B., Rizzo, G.: A model for evaluating the environmental and functional benefits of "innovative" roundabouts. Transp. Res. Part D Transp. Environ. **39**, 1–16 (2015). https://doi.org/10.1016/j.trd.2015.05.004
11. Zhai, Z., Tu, R., Xu, J., Wang, A., Hatzopoulou, M.: Capturing the variability in instantaneous vehicle emissions based on field test data. Atmosphere **11**(7), 765 (2020)
12. Abo-Qudais, S., Qdais, H.A.: Performance evaluation of vehicles emissions prediction models. Clean Technol. Environ. Policy **7**(4), 279–284 (2005)
13. Jiménez-Palacios, J.L.: Understanding and quantifying motor vehicle emissions with vehicle specific power and TILDAS remote sensing. Ph.D. Thesis, Massachusetts Institute of Technology, Department of Mechanical Engineering (1999)
14. Bishop, J.D.K., Stettler, M.E.J., Molden, N., Boies, A.M.: Engine maps of fuel use and emissions from transient driving cycles. Appl. Energy **183**, 202–217 (2016)
15. Frey, H.C., Zhang, K., Rouphail, N.M.: Fuel use and emissions comparisons for alternative routes, time of day, road grade, and vehicles based on in-use measurements. Environ. Sci. Technol. **42**(7), 2483–2489 (2008)
16. Vasconcelos, L., Silva, A.B., Seco, Á.M., Fernandes, P., Coelho, M.C.: Turbo roundabouts multicriterion assessment of intersection capacity, safety, and emissions. Transp. Res. Rec. **2402**, 28–37 (2014)
17. Fernandes, P., Pereira, S.R., Bandeira, J.M., Vasconcelos, L., Bastos Silva, A., Coelho, M.C.: Driving around turbo-roundabouts vs. conventional roundabouts: Are there advantages regarding pollutant emissions? Int. J. Sustain. Transp. **10**(9), 847–860 (2016).
18. AIMSUN Dynamic Simulator User Manual 8, Transport Simulation System (TSS) (2011)
19. Giuffrè, O., Granà, A., Giuffrè, T., Acuto, F., Tumminello, M.L.: Estimating pollutant emissions based on speed profiles at urban roundabouts: a pilot study. In: Sierpiński, G. (ed.) TSTP 2019. AISC, vol. 1091, pp. 184–200. Springer, Cham (2020). https://doi.org/10.1007/978-3-030-35543-2_15
20. Ministero delle Infrastrutture e dei Trasporti [Ministry of Infrastructure and Transport]. http://www.mit.gov.it/mit/mop_all.php?p_id=13799. Accessed 14 Apr 2021
21. Roundabouts - Application and design. A practical manual Royal Askoning. Ministry of Transport, Public Works and Water management Partners for Roads June (2009)
22. Dodi, D.: Il concetto di turbo rotatoria. Un'applicazione progettuale [The turbo roundabout concept. A design application] Bachelor's Thesis in Urbanistic and City Science, University of Palermo, Italy (2019)
23. Coelho, M.C., Farias, T.L., Rouphail, N.M.: Effect of roundabout operations on pollutant emissions. Transp. Res. Part D Transp. Environ. **11**(5), 333–343 (2006)
24. Bokare, P.S., Maurya, A.K.: Acceleration-deceleration behaviour of various vehicle types. Transp. Res. Procedia **25**, 4733–4749 (2017)
25. USEPA. Methodology for developing modal emission rates for EPA's multiscale motor vehicle and equipment emission system (EPA420). North Carolina State University for USEPA, Ann Arbor, MI (2002)

26. Coelho, M.C., Frey, C.H., Rouphail, N.M., Zhai, H., Pelkmans, L.: Assessing methods for comparing emissions from gasoline and diesel light-duty vehicles based on microscale measurements. Transp. Res. Part D **14**,91–99 (2009)
27. Pronello, C., Andre, M.: Speed and Acceleration Impact on Pollutant Emissions, SAE Technical Paper Series, International Spring Fuels & Lubricants Meeting Dearborn, Michigan, 6–8 May 1996 (1996)
28. Anya, A.R., Rouphail, N., Frey, H.C., Schroeder, B.J.: Application of AIMSUN microsimulation model to estimate emissions on signalized arterial corridors. Transp. Res. Rec. J. Transp. Res. Board **2428**(2428), 75–86 (2014)
29. Barceló, J.: Fundamentals of Traffic Simulation. Springer, London (2010). https://doi.org/10.1007/978-1-4419-6142-6
30. Giuffrè, O., Granà, A., Giuffrè, T., Acuto, F., Lo Pinto, A.: Life-cycle costing decision-making methodology and urban intersection design: modelling and analysis for a circular city. In: Macioszek, E., Sierpiński, G. (eds.) Research Methods in Modern Urban Transportation Systems and Networks. LNNS, vol. 207, pp. 59–86. Springer, Cham (2021). https://doi.org/10.1007/978-3-030-71708-7_5

# Use of ITS Data in Planning and Organization of Traffic Flows

# Geo-Information Input Data for the Planning Process of Conventional Fleet Conversion towards a 100% Electric Bus Fleet

Marcin Jacek Kłos(✉) and Stanisław Krawiec

Faculty of Transport and Aviation Engineering, Silesian University of Technology, Katowice, Poland
{marcin.j.klos,stanislaw.krawiec}@polsl.pl

**Abstract.** European environmental and transport policies commit, among others, to increasing the number of zero-emission vehicles in cities and agglomerations. Strategic documents at the national level also set out similar commitments. One of the available options is the increasing use of electric buses. This requires a transition to a fleet consisting at least in part of electric buses. Conducting a conversion process to restructure a public transport bus fleet without proper planning is often a difficult task. The process of such conversion is usually long and should be well planned in technical, economic and organizational dimensions. This paper presents selected management dilemmas of the planning process and a set of tools to help in planning the conversion process of a city bus fleet. It also discusses the influence of geo-information input data on the proper functioning of IT applications.

**Keywords:** Public transport · Electromobility deployment · Electric bus · GIS · Spatial analysis

## 1 Introduction

The intense social pressure on environmental aspects of transport and endeavours to sustainable mobility requires an increase in the number of trips with the use of public transport [1–5]. It is caused having regard to the provisions of the European Union's policy on transport [6, 7]. The aim of reducing the nuisance of transport in urban areas requires public transport operators to take actions and decide in terms of replacing their bus fleet with a more ecological one [8, 9]. Battery electric buses (electric buses hereafter) are one of the technologies available [10–14]. The problem of bus fleet transition towards a fully electric bus fleet is a challenge that many cities in Europe face [15–18]. This process is considered complicated, multithreaded and multifaceted [10, 19, 20].

Due to the high investment costs of the investment and organizational difficulties, the bus fleet transition process may either be implemented gradually or at once. Under existing limitations, most cases cannot do that without planning such a process properly. Hence, there is a need to develop professional tools to support this process. Such is a PLATON Toolkit. This paper presents its functionalities and an analysis of the useful geo-information data needed to support the bus fleet conversion process. We also show


G. Sierpiński (Ed.): TSTP 2021, LNNS 352, pp. 49–60, 2022.
https://doi.org/10.1007/978-3-030-91156-0_4

the examples of the data with the visualization based on the Metropolitan Association of Upper Silesia and Dąbrowa Basin (officially abbreviated to 'GZM Metropolis') area, located in southern Poland [21].

## 2 PLATON Toolkit's Components to Support the Fleet Conversion Planning Process

The planning process for the conversion of conventionally powered buses (diesel, hybrid) to electric propulsion takes place at the management level of public transport operators and transit authorities. The goal of this process is to reach a point where as much as possible, ideally 100% of the bus fleet, is electrically powered [22]. A one-off deployment of electric buses for all feasible public vehicle cycles seems almost always impossible, so the planning and replacement process is often carried out in stages, as explained below.

Until electric buses meeting the requirement of the appropriate range in all-weather conditions are developed (which would enable simple one-by-one replacement) [11, 12, 23, 24], three strategic options should be considered to increase the share of electric buses in the fleets of public transport operators. The first option assumes maintaining the existing transport offer, keeping the structure of vehicle cycles unchanged. The second option is to modify the assignment of buses to bus routes with no changes in timetables for passengers. A partial or complete change in the bus schedule is considered in option three, which however implies the need to change in vehicle cycles. These approaches can be complemented by the following long-term strategies for the bus fleet transition process [25]:

- An offensive strategy that assumes one-off replacement of the existing bus fleet with electric buses, regardless of previous operating experience
- A cautious strategy which assumes testing several electric buses to gain exploitation experience, and afterwards a decision to replace the remaining fleet
- A step-by-step conversion strategy on the way to the highest possible share of electric buses
- A defensive strategy in which the moment of fleet replacement is postponed to an unspecified future, waiting for experiences of other entities.

Presumably, most transport companies will follow the step-by-step conversion strategy. Such action is justified for economic grounds, especially from the total cost of ownership (TCO) analysis [26–29]. Nevertheless, all of the above-mentioned strategies may and should be supported by information technology. Such is a toolkit developed within the project PLATON. It includes 11 components that have been developed by an international project consortium from Germany, Poland, Austria and Belarus [30]. These are [31, 32]:

- DataProc that enables the generation of trip cycles in the form of a speed plan for a given route in a public transport network
- CellParameters to build battery models

- CollectApp that allows real-time data collection about the potential route to be operated by electric buses
- SyntheticalTrips enables a simulation model for the automatic generation of speed plans for a hypothetical route in a transit network.

The role of the above four components is to provide input data for the Bus vehicle simulation [33] component, which aims to support public transport operators in providing reliable energy efficiency calculations for a given trip on a real route to determine the feasibility of equipping the bus with batteries of the appropriate capacity. The component can also be used to investigate the electrification priority.

Next, we have another six components that support the selected elements of the bus fleet conversion process [34, 35]:

- ECBus + allows calculating the energy consumption of an electric and diesel bus on a given route planned for electric buses when speed profile, bus and road network parameter data given
- OptimSched is designed to solve optimization problems related to the planning of the fleet conversion process and includes three optimization modules: Opt, DepOpt and OptSched
- VisualGrids provides public transport operators with a web-based tool to search and edit bus routes for studying their possible electrification. This component is a web application based on OpenStreetMap using Overpass API
- NMEA simulator, which allows calculation of the energy balance for the operating models (charging types) tested
- TCOModel enables the calculation of the TCO of electric buses by the user over their entire life cycle. This includes the investment costs for operating a fleet of electric buses + supporting infrastructure + electricity over its entire life cycle. This component involves using the Present Value (PV) method to estimate the present value of future cash flows. The TCOModel component provides a range of options for financing the investment, taking into account various possibilities and combinations: own funds, bank loan, subsidies (grants) or leasing.
- ReportGenerator that generates a reporting document.

All components and functionalities of the project are publicly available online [36]. They are intended for the management of transport companies (including public transport operators), local authorities (e.g. municipalities, regions) involved in transport planning in a given area and public transport organisers).

Efficient and reasonable use of the PLATON application requires preparation of the following input data sets:

- Market data on current and forecast prices of electric energy, costs of electric buses, costs of charging infrastructure and other infrastructure necessary for the operation of electric buses
- Data from the business environment on financial validity of long term investment processes planning

- Data on technical parameters of electric buses and necessary infrastructure for their operation, including parameters necessary to manage electric safety
- Data characterizing the facility for which calculations will be conducted, including data on the transportation network operated or intended to be operated by electric buses.

The above-mentioned input data necessary for the proper operation of the PLATON toolkit also take advantage of other widely available applications based on geo-information data.

## 3 Geo-Information Data Useful in the Fleet Conversion Planning Process

The basis for the planning process is input data with the current status of the public transport system being considered. The local landform (elevation and depression), size of the area, number of public transport lines, road network and the availability of charging infrastructure served by the operation of individual operators may significantly differ from each other. The use of multiple-input data (e.g. geoinformation data) allows for the implementation of the described analysis. We may take these data using geoinformation technology as it allows describing the current state of the transport network in the analysis area. Depending on the country, the range of data available may be different. The development of a universal IT tool needs consideration of available data in the countries analyzed.

There are many types of format files existing in the GIS system. The most popular formats are [37]:

- Shapefiles shape format (SHP),
- Data base file (DBF),
- Shapefile shape index format (SHX),
- Projection description (PRJ),
- Comma-separated values (CSV),
- Keyhole markup language (KML),
- Geographical JavaScript object notation (GeoJSON).

A General Transit Feed Specification (GTFS) format is becoming more popular as a format for keeping bus schedules. This type is defined as a standard format for public transportation timetables associated with spatial information. For that case, we prepare an essential solution for the final data format, and we did more accurate analysis for GTFS format because many countries use such of the type of files [38]. Some European countries use their type of format: Transmodel, which contains timetable and network data [39]. Germany and German-speaking countries use another type: VDV 457 Standard [40].

**Table 1.** Bus stops – description of the data format.

| Data | Example of the data |
|---|---|
| ID | 1 |
| Name | Sosnowiec Osiedle Piastów |
| Longitude | 19.112192 |
| Latitude | 50.279738 |
| Type (standard or on-demand bus stop) | Standard |
| Number of lines | 8 |
| Name of the lines | 154, 299, 808, 811, 813, 815, 835, 908N |
| Line frequency | Link to another table |
| Altitude [m] | 253 |
| Charger [Yes or No] | No |

We prepared two main tables with data for the bus stops and inter-stops sections.

Table 1 presents the data format for the bus stops. The basic concept is the representation of the bus stops as a point on the map. Bus stops are identified by the directions of driving and with the stands. One bus stop could contain many stands, especially at larger bus stations.

The following table presents exemplary data formats for the inter-stops sections. These contain the length and the information on whether in-motion charging is available.

**Table 2.** Inter-stop sections – description of the data format.

| Data | Example of the data |
|---|---|
| ID | 1 |
| First bus stop | Gliwice Świętojańska |
| Second bus stop | Szobiszowice |
| Length [km] | 0.950 |
| In-motion charging [Yes or No] | No |

Data about all bus stops stands and inter-stop sections allow conducting calculations. The data from Tables 1 and 2 allow analyzing the routes of all bus lines on the transportation network. Geo-information data with length and altitude permit us to prepare height adjustment profiles, which would help calculate energy consumption. Location of the chargers at bus stop and identification of the sections with charging in motion could also be input data for the IT tool.

## 4 Sources of the GIS Data

Depending on the country or city, there are significant variations in the GIS data availability and these data sources. Examples of the source of the data could be OpenStreetMap (OSM) [41], GPS data from buses [42] or information prepared by buses operator (e.g. GTFS). Those three sources of data of public transport are the most popular in Europe [43].

OSM data are freely available for research purposes. Those data were recorded in the specific form using own tags (all types of tags are explained on the OSM website) [44]. Users use specific tags by a query to obtain data connected with public transport. Figure 1 shows the example of the results of the query to the OSM website. The description of the presented tags is the following:

- From – the start of the route from the displayed bus stop,
- Name – the name of the bus line and the relation (first bus stop and destination),
- Network – the operator of the bus,
- Ref – the number of the bus,
- Route – the type of the public transport vehicles,
- To – the destination bus stop,
- Type – the type of the bus line.

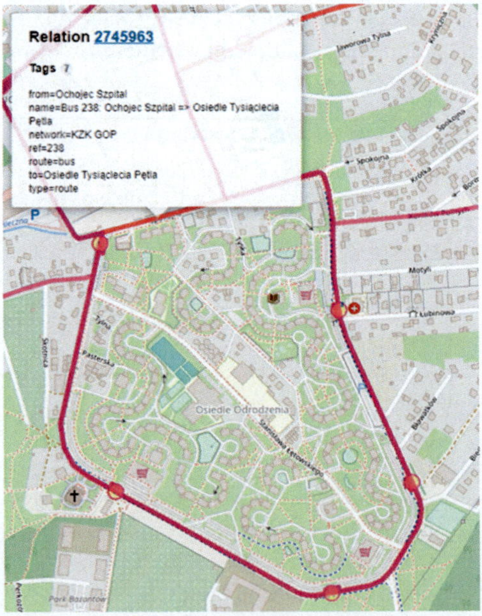


**Fig. 1.** Example of the queries to the OSM website (Source: Map data copyrighted OpenStreetMap contributors and available from https://www.openstreetmap.org)

Figure 1 contains the information mentioned above (tags) and the visualization of the route with the bus stops. This kind of data could be export to the SHP or CSV file, which is helpful to pre-processing data to the form proposed in section two.

The following typical source of data from public transport is the recorded data from the GPS installed in the buses. Recorded data could have a different recording frequency [45]. Analyzed data (from the SM area) has a 6-s recording frequency. This kind of data has all basic data and contains additional information, which could be helpful in other research. Those data typically have information about (based on SM data):

- Line – information about the number of buses,
- Bus stop – identification of all bus stops for the analyzed line,
- Previous/next bus stop – information about previous and next bus stop for the analyzed line (those data could be compared with raw location data of the bus stops)
- Time – data with time of one GPS record,
- Longitude, latitude – representation of the spatial with X and Y,
- Speed – instantaneous vehicle speed,
- Vehicle ID – a unique number of the bus.

That additional information of the GPS data could be useful for calculating energy consumption. The knowledge about the unique vehicle number makes it possible to obtain vehicle cycles for every bus. That data allows calculating energy consumption for whole tasks. Typical timetable information (available for the passengers) does not contain data about all vehicle tasks e.g., road from and depot or technical breaks. The GPS data contain all of that information (from the start of the bus) to extend the energy consumption calculation. Figure 2 shows the visualization of GPS data from one route of the SM area. Recorded data from GPS mounted in the bus have points representation. In that case, one point is generated every 6 s.

The last source of GIS data depends on the bus operator, which prepares this information. Various operators may have internal systems for preparing timetables and offer for passengers. Another problem with such data is making them available to the public, as different countries and even cities have divergent policies in this regard. (For example, the Metropolitan Transport Authority of the Metropolis GZM shares such data freely on the internet in GTFS format). The format construction is based on 9 files, which are presented in Table 3.

The problem with such data is the operator's uploading because depending on the country or even city, such files could be incomplete, e.g., without shapes.txt (it is not possible to create real paths of the buses). This causes to not be obtained with automatic plugin altitude profile of the path, which has a significant impact on the calculation of electric bus energy uses.

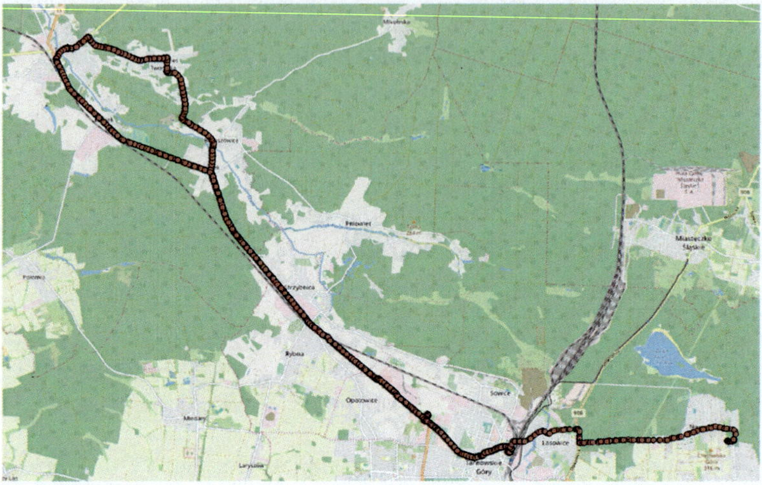

**Fig. 2.** Example of the GPS data in the GZM Metropolis area (Source: Map data copyrighted OpenStreetMap contributors and available from https://www.openstreetmap.org)

**Table 3.** GTFS construction files.

| Name of the files | Content |
|---|---|
| Agency.txt | Information about agencies, which realize public transports |
| Calendar.txt | General information about dates (workdays/days off/holidays) |
| Calendar_dates.txt | Accurate dates and time for the current version of GTFS files |
| Feed_info.txt | Information about the organisation, which shares the data |
| Routes.txt | Information about the routes of the buses (with a description of all bus stops) |
| Shapes.txt | The data, which allows preparing tracks for buses on the road network |
| Stop_times.txt | Timetables information connected with the bus stops |
| Stops.txt | Information about all bus stops with stands |
| Trips.txt | Data of all trips connected with bus stops and shapes.txt |

Figure 3 presents an example of the visualization of the GTFS data from the GZM Metropolis area. Red points represent the data from shapes.txt files – nodes of the routes of the buses. Blue markers are the bus stops. Blackline is the pre-processing effect of the connecting points from shapes.txt, showing the bus's real route.

Analysis of the source of the input data for the PLATON toolkit is necessary for the context of the feature usage. Presented sources are most popular in the European country. Depending on the country, there is a need to making pre-processing the raw data.

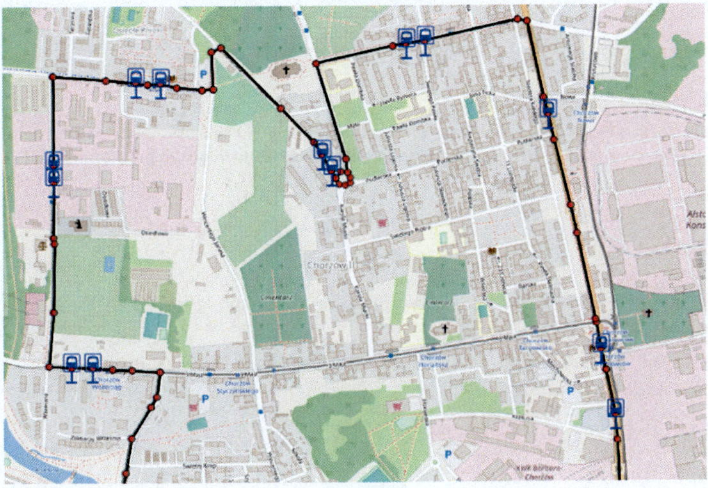

**Fig. 3.** Example of the GTFS data visualization (Source: Map data copyrighted OpenStreetMap contributors and available from https://www.openstreetmap.org)

## 5 Conclusion

The fleet conversion planning process requires interrelated decisions made in the context of what is currently going on in the transport systems. The dynamics of these processes and the often multi-year planning horizon force the planning process to be highly adaptive. The political will, the ability to overcome external and internal conditions, and the consideration of technical and economic issues create a gap in many possible strategies and activities. Therefore, the conversion process has to be supported by experts and information technology. Functionalities of the PLATON Toolkit components have been presented in the article. This toolkit enables multidimensional support of the planning process of conversion of the traditional fleet of public transport buses in cities and agglomerations to a fleet consisting as much as possible of electric buses. The paper also presents geo-information data sources and files necessary to carry out the conversion process and indicates differences in available data format. The presented data is the primary input for all PLATON Toolkit. With the application of such data, users could e.g. calculate energy consumption on the route (with speed profile and road network parameter), redefine routes for electrification, calculate TCO.

**Acknowledgement.** The present research has been financed from the means of the National Centre for Research and Development as a part of the international project within the scope of ERANET EMEurope programme 'Planning Process and Tool for Step-by-Step Con-version of the Conventional or Mixed Bus Fleet to a 100% Electric Bus Fleet.

The project "Planning Process and Tool for Step-by-Step Conversion of the Conventional or Mixed Bus Fleet to a 100% Electric Bus Fleet" has been financed from the means of the National Centre for Research and Development (Poland) within the scope of ERANET EMEurope programme.

## References


1. Banister, D.: The sustainable mobility paradigm. Transp. Policy. **15**, 73–80 (2008). https://doi.org/10.1016/j.tranpol.2007.10.005
2. Köhler, J., Whitmarsh, L., Nykvist, B., Schilperoord, M., Bergman, N., Haxeltine, A.: A transitions model for sustainable mobility. Ecol. Econ. **68**, 2985–2995 (2009). https://doi.org/10.1016/j.ecolecon.2009.06.027
3. Mugion, R.G., Toni, M., Raharjo, H., Di Pietro, L., Sebathu, S.P.: Does the service quality of urban public transport enhance sustainable mobility? J. Clean. Prod. **174**, 1566–1587 (2018). https://doi.org/10.1016/j.jclepro.2017.11.052
4. Wang, L., Xue, X., Zhao, Z., Wang, Z.: The impacts of transportation infrastructure on sustainable development: emerging trends and challenges. Int. J. Environ. Res. Public Health. **15**, 1172 (2018). https://doi.org/10.3390/ijerph15061172
5. Boulange, C., Gunn, L., Giles-Corti, B., Mavoa, S., Pettit, C., Badland, H.: Examining associations between urban design attributes and transport mode choice for walking, cycling, public transport and private motor vehicle trips. J. Transp. Heal. **6**, 155–166 (2017). https://doi.org/10.1016/j.jth.2017.07.007
6. WHITE PAPER Roadmap to a Single European Transport Area – Towards a competitive and resource efficient transport system. European Commission (2011)
7. EU climate action and the European Green Deal | Climate Action. https://ec.europa.eu/clima/policies/eu-climate-action_en. Accessed 21 Apr 2021
8. Xylia, M., Silveira, S.: On the road to fossil-free public transport: the case of Swedish bus fleets. Energy Policy **100**, 397–412 (2017). https://doi.org/10.1016/j.enpol.2016.02.024
9. Taczanowski, J., Kołoś, A., Gwosdz, K., Domański, B., Guzik, R.: The development of low-emission public urban trans-port in Poland. Bull. Geogr. Socio-economic Ser. **41**, 79–92 (2018). https://doi.org/10.2478/bog-2018-0027
10. Biresselioglu, M.E., Demirbag Kaplan, M., Yilmaz, B.K.: Electric mobility in Europe: a comprehensive review of motivators and barriers in decision making processes. Transp. Res. Part A Policy Pract. **109**, 1–13 (2018). https://doi.org/10.1016/j.tra.2018.01.017
11. Göhlich, D., et al.: Design of urban electric bus systems. Des. Sci. 4 (2018). https://doi.org/10.1017/dsj.2018.10.
12. Li, X., Castellanos, S., Maassen, A.: Emerging trends and innovations for electric bus adoption—a comparative case study of contracting and financing of 22 cities in the Americas, Asia-Pacific, and Europe. Res. Transp. Econ. **69**, 470–481 (2018). https://doi.org/10.1016/j.retrec.2018.06.016
13. Mahmoud, M., Garnett, R., Ferguson, M., Kanaroglou, P.: Electric buses: a review of alternative powertrains (2016). https://doi.org/10.1016/j.rser.2016.05.019.
14. Nanaki, E.A., et al.: Environmental assessment of 9 European public bus transportation systems. Sustain. Cities Soc. **28**, 42–52 (2017). https://doi.org/10.1016/j.scs.2016.08.025
15. Malnaca, K., Yatskiv (Jackiva), I.: Impact of critical variables on economic viability of converted diesel city bus into electric bus. In: Nathanail, E.G., Karakikes, I.D. (eds.) CSUM 2018. AISC, vol. 879, pp. 847–855. Springer, Cham (2019). https://doi.org/10.1007/978-3-030-02305-8_102

16. Pelletier, S., Jabali, O., Mendoza, J.E., Laporte, G.: The electric bus fleet transition problem. Transp. Res. Part C Emerg. Technol. **109**, 174–193 (2019). https://doi.org/10.1016/j.trc.2019.10.012
17. Deliali, A., Chhan, D., Oliver, J., Sayess, R., Godri Pollitt, K.J., Christofa, E.: Transitioning to zero-emission bus fleets: state of practice of implementations in the United States. Transp. Rev. (2020). https://doi.org/10.1080/01441647.2020.1800132
18. Liu, D., Buley, J.: The US electric bus transition: an analysis of funding and financing mechanisms. (2019)
19. Bakker, S., Konings, R.: The transition to zero-emission buses in public transport – the need for institutional innovation. Transp. Res. Part D Transp. Environ. **64**, 204–215 (2018). https://doi.org/10.1016/j.trd.2017.08.023
20. Islam, A., Lownes, N.: When to go electric? A parallel bus fleet replacement study. Transp. Res. Part D Transp. Environ. **72**, 299–311 (2019). https://doi.org/10.1016/j.trd.2019.05.007
21. Metropolia GZM. https://metropoliagzm.pl/en/. Accessed 22 Apr 2021
22. Krawiec, K.: A concept of conventional or mixed bus fleet conversion with electric vehicles: a planning process. Int. J. Electr. Electron. Eng. Telecommun. **9**(1), 8–12 (2020).https://doi.org/10.18178/ijeetc.9.1.8-12
23. Borén, S.: Electric buses' sustainability effects, noise, energy use, and costs. Int. J. Sustain. Transp. **14**, 956–971 (2020). https://doi.org/10.1080/15568318.2019.1666324
24. Varga, B.O., Iclodean, C., Mariasiu, F.: Electric and Hybrid Buses for Urban Transport. Springer, Cham (2016). https://doi.org/10.1007/978-3-319-41249-8
25. Krawiec, S., et al.: Urban public transport with the use of electric buses – development tendencies. Transp. Probl. **11**, 127–137 (2016). https://doi.org/10.20858/tp.2016.11.4.12
26. Goswami, R., Tripathi, G.C.: Economic, environmental and congestion impact on the life-cycle cost of ownership: a case study in the Delhi transit bus system. Int. J. Electr. Hybrid Veh. **11**(1), 59–72 (2019)
27. Pagliaro, M., Meneguzzo, F.: Electric bus: a critical overview on the dawn of its widespread uptake. Adv. Sustain. Syst. **3**, 1800151 (2019). https://doi.org/10.1002/adsu.201800151
28. Krawiec, S., et al.: Economic conditions to introduce the battery drive to busses in the urban public transport. Transp. Res. Proc. **14**, 2630–2639 (2016). https://doi.org/10.1016/j.trpro.2016.05.420
29. Grauers, A., Borén, S., Enerbäck, O.: Total cost of ownership model and significant cost parameters for the design of electric bus systems. Energies **13**, 3262 (2020). https://doi.org/10.3390/en13123262
30. Czogalla, O., Xie, F.: Electric bus deployment tools for strategic and operational planning stages. In: 27th World Congress on Intelligent Transport Systems, Los Angeles (2020)
31. Xie, F., Czogalla, O., Shi, H.: Application of lithium-ion battery thermal management system in electric vehicle simulation. In: Krawiec, K., Markusik, S., Sierpiński, G. (eds.) Electric Mobility in Public Transport—Driving Towards Cleaner Air. LNITI, pp. 135–147. Springer, Cham (2021). https://doi.org/10.1007/978-3-030-67431-1_9
32. Czogalla, O., Jumar, U.: Design and control of electric bus vehicle model for estimation of energy consumption. In: IFAC-PapersOnLine, pp. 59–64. Elsevier B.V. (2019). https://doi.org/10.1016/j.ifacol.2019.12.381.
33. Czogalla, O., Xie, F.: Methods and tools for public bus fleet electrification in the area of sustainable city transportation. In: 26nd World Congress on Intelligent Transport Systems, pp. 21–25, Singapore (2019)
34. Kovalyov, M.Y., Rozin, B.M., Guschinsky, N.N.: Mathematical model and random search algorithm for the optimal planning problem of replacing traditional public transport with electric. Autom. Remote Control. **81**, 803–818 (2020). https://doi.org/10.1134/S0005117920050033

35. Algin, V., Goman, A., Skorokhodov, A., Bytsko, O., Chistov, S., Fedasenka, S.: Methodology for probabilistic assessment of energy consumption by electric buses on routes. In: Krawiec, K., Markusik, S., Sierpiński, G. (eds.) Electric Mobility in Public Transport—Driving Towards Cleaner Air. LNITI, pp. 83–105. Springer, Cham (2021). https://doi.org/10.1007/978-3-030-67431-1_6
36. PLATON project website. http://projektplaton.polsl.pl/. Accessed 22 Apr 2021
37. Ng, C.P., Law, T.H., Jakarni, F.M., Kulanthayan, S.: Relative improvements in road mobility as compared to improvements in road accessibility and urban growth: A panel data analysis. Transp. Res. Part A Policy Pract. **117**, 292–301 (2018). https://doi.org/10.1016/j.tra.2018.08.032
38. Jakimavičius, P.: Antuchevičiene, Karpavičius: internet GIS-based multimodal public transport trip planning information system for travelers in Lithuania. ISPRS Int. J. Geo-Inf. **8**, 319 (2019). https://doi.org/10.3390/ijgi8080319
39. Benvenuti, F., Diamantini, C., Potena, D., Storti, E.: An ontology-based framework to support performance monitoring in public transport systems. Transp. Res. Part C Emerg. Technol. **81**, 188–208 (2017). https://doi.org/10.1016/j.trc.2017.06.001
40. Kaufmann, S., Ulm, U., für Ingenieurwissenschaften, F.: Opening public transit data in Germany.
41. Mooney, P., Minghini, M.: A review of OpenStreetMap data. https://doi.org/10.5334/bbf.c.
42. Garg, N., Ramadurai, G., Ranu, S.: Mining bus stops from raw GPS data of bus trajectories. In: 2018 10th International Conference on Communication Systems and Networks, COMSNETS 2018, pp. 583–588. Institute of Electrical and Electronics Engineers Inc. (2018). https://doi.org/10.1109/COMSNETS.2018.8328278.
43. Kłos, M.J., Sierpiński, G.: Building a model of integration of urban sharing and public transport services. Sustainability **13**(6), 3086 (2021). https://doi.org/10.3390/su13063086
44. Żochowska, R., Jacyna, M., Kłos, M.J., Soczówka, P.: A GIS-based method of the assessment of spatial integration of bike-sharing stations. Sustainability **13**(7), 3894 (2021). https://doi.org/10.3390/su13073894
45. Pashkevich, A., Kłos, M.J., Jaremski, R., Aristombayeva, M.: Method to evaluate a bike-sharing system based on performance parameters. In: Sierpiński, G., Macioszek, E. (eds.) Decision Support Methods in Modern Transportation Systems and Networks. LNNS, vol. 208, pp. 95–113. Springer, Cham (2021). https://doi.org/10.1007/978-3-030-71771-1_7

# Information Interaction in Seaports and Port Stations

Sergey Vakulenko[1], Peter Kurenkov[1](✉), Grigoriy Deruzhinskiy[2], Alexander Ignatenko[3], Vladimir Nekhoroshkov[4], Alyona Aroshidze[4], Nikita Moisievich[5], and Irina Seryapova[6]

[1] Institute of Management and Digital Technologies, Russian University of Transport (RUT–MIIT), 9, Building 9, Obraztsova Street, Moscow 127994, Russia
post-iuit@bk.ru

[2] Faculty of Water Transport Operation and Navigation, Admiral F. F., Ushakov State Maritime University, 93, Lenin's Avenue, Novorossiysk 353918, Russia
evropa@bk.ru

[3] Marine Engineering Faculty, Admiral F. F., Ushakov State Maritime University, 93, Lenin's Avenue, Novorossiysk 353918, Russia

[4] Faculty International Business and Law, Siberian Transport University, 191, Dusi Kovalchuk Street, Novosibirsk 630049, Russia

[5] Directorate of Terminal and Warehouse Complexes, Administration of the Kuibyshev Railway, 2, Komsomolskaya Square, Samara 443030, Russia
nikita_vm@inbox.ru

[6] Faculty of Economics, Logistics and Management, Samara State University of Railway Transport, 2V, Svobody Street, Samara 443008, Russia

**Abstract.** The presented article examines the problems that arise in the interaction of subjects of the transport and logistics process and suggests ways to solve them. The existing process of interaction of modes of transport in transport hubs is considered, which does not meet the modern requirements of subjects of interaction, in connection with this situation, an analysis of information integration into transport services in ports and port stations is carried out. The latest information technologies make it possible for the subjects of interaction to work in a single information field, which guarantees the relevance, integrity and consistency of information that will be used to make the necessary management decisions by all participants in the transportation process.

Approbation and economic justification of measures based on the implementation of an information system at the port station in cooperation with the port showed a positive economic effect. The developed methodology is focused on determining the information integration effectiveness for transport services in ports and port stations based on the organization of a common or unified information field. The introductory part of the article describes the traditional and modern systems of information interaction. The second part examines the evolution of information interaction between seaports and the railway. The second part also considers the outcomes of Road Information and Logistics System (DILS) implementation as well. The third part presents the results of the economic effect calculations based on information interaction improvements and vehicle downtime reduction .


G. Sierpiński (Ed.): TSTP 2021, LNNS 352, pp. 61–71, 2022.
https://doi.org/10.1007/978-3-030-91156-0_5

**Keywords:** Interaction · Transport · Logistics · Carrying capacity · Unloading area · Port · Information system · Rail and water communication · Port station

## 1 Introduction

Recently, an increase in the number of transshipment cargo from rail to water transport in ports has revealed the problems, inefficiency of interaction between water and railway enterprises. These problems are expressed in excess of the standard idleness of wagons, which is still far from optimistic forecasts [1].

The North Caucasian Railway serves the Russian port network in the Azov and Black Seas.

The previously developed unified technological processes for the implementation of multimodal transportation, continuous schedules of interaction between enterprises at the units and other managerial decisions are not always effective and relevant for the prevailing conditions during the transportation process [2, 3]. The method of managing the supply of goods to ports does not meet the modern requirements of consignees and the port in the optimization aspect of the choice of service tools, since in many transport hubs today, there is a standard for the interaction of the port station and the port, based on the transfer of information trough e-mail, telephone and other means of communication, not connected to a single digital space.

The methodology for managing the supply of goods to ports that still exists in many transport hubs is shown in Fig. 1.

At the present stage of development, in order to ensure the increasing volumes of transportation, the development and implementation of the Concept of logistics management is being carried out. The fundamental element of the new Concept was the creation of conditions for information interaction between the structural divisions of Russian Railways and seaport operators. The creation of such conditions is aimed at improving the quality of forecasting and planning, minimizing unproductive losses of railway stations and port terminals, reducing the downtime of the fleet and transshipment equipment due to the misuse of the station and port infrastructure [4].

In order to improve the situation in the problem area, new information technologies called "Road Information and Logistics System" (DILS) were developed and introduced. These technologies allow achieving the optimal rhythm in the process of promotion and uniformity of loading of terminals at the point of repayment of the flow of goods either at the station or in the port.

Research in the existing methods for the efficiency of information integration for transport services in ports and port stations is based on the introduction of updated information technologies associated with the need to organize a common information field. This, in turn, guarantees the relevance, integrity and consistency of information that will be used to make the necessary management decisions by all parties involved in the transportation process, taking into account the fact that each of the participants has their own automated system [5, 6].

This interaction of railway and water transport on the basis of the presented information system will make it possible to increase the profitability of both the port station

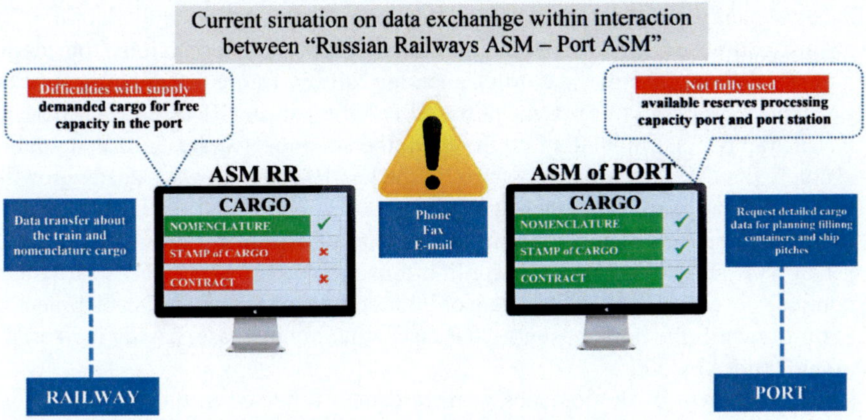


**Fig. 1.** Methodology for managing cargo deliveries to ports

and the port, saving the port station from idle time of loaded trains, while ships free of cargo are standing in the port.

Analysis of modern scientific literature on the issues under study showed that the theoretical aspects of information integration of transport services in ports and port stations fell into the field of scientific research of such scientists as Deitz G., Zubkov E., Kuzina E., Liu G., Makeev V. Chislov O.N. Mamaev E.A. and others. However, the studies in the field of the information integration efficiency for transport services in ports and port stations were carried out extremely rarely.

The scientific novelty of the research covers the improvements and benefits of information integration for transport services in ports and port stations based on the economic efficiency related to the implementation of the DILS system at the port station. The proposed methodology to assess the economic efficiency of "DILS" project implementation counts on positive dynamics in key performance indicators relevant to railway port stations and ports interaction, in particular – the idle time reduction for the wagons at the port station. The theoretical significance is shown through the complex solution development – ensuring the effectiveness of information integration of transport services in ports and port stations, covering various modes of transport.

The practical significance of the study is manifested in the use of the developed tools in the process of determining the effectiveness of information integration of transport services in ports and port stations, as well as substantiating the management decisions taken.

## 2 Description of the Approach to Assessing the Effectiveness of Information Interaction between Port Stations and Ports Based on the Implementation of the "Kontur DILS" System

Researches have shown that one of the indicators of the quality of transportation is the rhythm of transportation, which influences transport enterprises when they interact in a transport hub [7]. Optimization of the above-mentioned indicator is possible through the logistics management, which is reflected in today's figures for the work of transport enterprises' entities during their interaction in rail-water communication. Considering, for example, the specifics of the North Caucasian railway traffic that mainly represents flows of exported cargoes towards the ports of the Azov Sea and Black Sea, it is observed in the structure of total unloading on the road, the share of wagons unloaded at above ports is more than 70%. And the dynamics of the last 10 years shows a steady growth in the transportation of export cargoes [8, 9].

Thus, unloading at the port stations of the North Caucasian Railway increased by more than 58% and amounted to 92 million tons in 2018. In 2019, the positive trend also continues. According to the Institute of Economics and Transport Development, the export flow towards the ports of southern Russia will only increase, having increased by more than 2 times by 2025.

The development of the logistics management is reflected in the current numbers of the ports. Along with an increase in unloading, in the past year all major ports have renewed their historical averages and highs. Novorossiysk in January 2018 reached the level of 2,043 wagons per day, Tuapse – 688 wagons and the maximum unloading on May, the 5$^{th}$ of 2020 – 1109 wagons, Vyshesteblievskaya – 469 wagons (maximum on the 16$^{th}$ of March 2020 – 1230 wagons), Grushevaya – 350 wagons (maximum in May 2017 – 630 wagons).

One of the main problems in ensuring the transport interaction of various kinds is work plans coordinating for all the participants in the transportation process. Export cargo arrives at the port mainly by rail and pipeline transport, where oil and oil products are mainly delivered via pipeline transport, and rail transport provides the bulk of the delivery of export cargo – coal, metals, fertilizers, timber and other cargo – 34,7% for the North Caucasian Railway. The seaport should smooth out the unevenness of traffic flows, but the limited port storage resources and the influence of meteorological factors (e.g. – strong winds and storms) put restrictions on flexibility in fulfilling this role. As a result, railway transport begins to play the role of a buffer in transportation, which leads to a decrease in its efficiency.

The ongoing modernization of the North Caucasian railway makes it possible to increase the unloading of wagons, but the problem cannot be solved only by extensive methods. To improve the interaction of modes of transport, information technologies of operational management are being introduced.

In the 1980s, it was considered necessary that the goods going for transshipment should be separated from the general loading; it was necessary to organize tracking of their movement on the communication network in order to inform the transport hubs in advance about the cargo flows directed to them. During this period, the exchange of information (movement of information flows) between the subjects was carried out in various ways and on various media – paper, speech, magnetic, via telegraph channel, electronic, as well as on diskettes, facsimile communication, etc. Information flows could move from one media to another: from paper to electronic (scanning of paper documents, manual input into a computer), from electronic to paper (printing on a computer printer of electronic documents received from a floppy disk or by e-mail), from speech to magnetic or vice versa (when recording on a tape recorder or listening to tape recordings) etc.

From the very beginning it became obvious that multiple exchange and transformations of data flows were slowing the process down due to the interaction speed limitations as well as due to additional errors introduced while information was transformed from one form to another.

When a large number of partners are involved in the exchange of information, standardization of the common database becomes imperative. Centralized control systems for the location of transport and cargo units are used for all modes of transport and transport departments of the exporters associated with carriers.

The data transmission networks created and developed in the period 1980–2010 in various modes of transport in Russia did not fully meet the requirements of electronic document management for various reasons.

The positive dynamics of unloading was largely achieved due to the introduction of the Road Information and Logistics System (DILS) on the North Caucasian Railway, which carries out information integration between the ports and Russian Railways.

The development of DILS on the North Caucasian railway is shown in Fig. 2.

The introduction of DILS on the North Caucasian railway will provide an opportunity for a unified digital control of cargo flows between port stations and the port. At the end of 2019, Novorossiysk Commercial Sea Port, Rostov Port, Tuapse Commercial Sea Port were connected to the DILS circuit.

Let us present the characteristics of the DILS information system in Fig. 3.

Based on the above, we believe that in the future, DILS should turn into a self-learning system with artificial intelligence.

The next direction in the development of information interaction between port stations and ports is to ensure further information exchange on the basis of the DILS system in the form of linking the automated system "Cargo Management" in the port to the "Automated local work management system" of JSC Russian Railways in the "ASM-ASM". This is necessary to ensure the transportation of increasing volumes of goods and is the introduction of a new concept of logistics management on the railway.

At the moment, the railway, and in particular the station, and the port "speak different languages." The railway controls the movement of trains to the port, and the port requires cargo of a certain nomenclature, brand and assortment. For example, for the railway "coal" is simply "coal", and for the port "coal" is DPK coal (long-flame fraction 50–100 mm), DPKO coal (long-flame, fraction 25–50 mm), TPK coal (lean, fraction 50–100 mm), etc.

Taking into account the fact that over the past ten years the nomenclature of goods arriving at the ports has increased significantly, and a division into assortments has appeared within the nomenclature, the task of drawing up a delivery plan taking into account the assortment position comes to the fore. Considering the volume of this task, its solution in manual mode is enlarged, which affects the quality of supplying the required assortment to the ports. The created module is designed to solve this problem.

The algorithm looks like this. Based on the data received from the port, according to the demand for cargo in the context of the assortment for the planned period, taking into account the availability of these goods on the road, a forecast of the arrival of trains with the required cargo is automatically generated 2 days ahead.

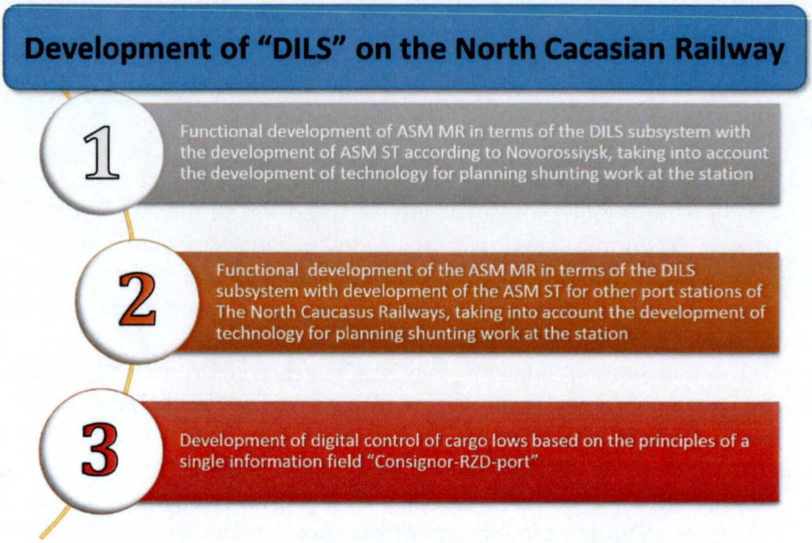


**Fig. 2.** Stages of implementation of "DILS" on the North Caucasian railway

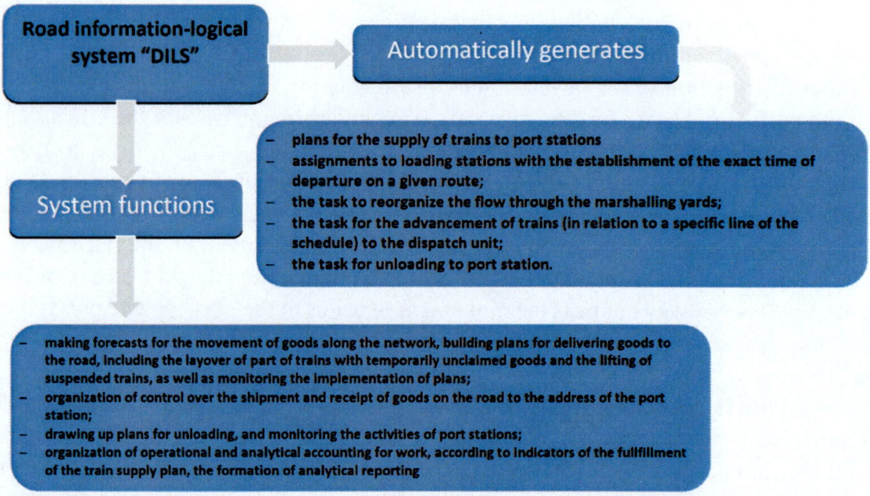


**Fig. 3.** Characteristics of the information system "DILS"

The use of this technology will ensure the rhythm in the movement and uniformity of loading of terminals at the point of repayment of the flow of goods. This, in turn, will have a positive effect on income, as well as on the cost of transportation of goods.

Let us present a methodology for calculating the economic efficiency from the implementation of the DILS system at port stations, taking into account that the economic effect of the introduction of automation tools can only be indirect, since the implemented automation tools are not a direct source of income, but are an auxiliary tool that minimizes costs.

## 3 Approbation and Economic Grounding of Measures from the Implementation of Information Systems at the Port Station in Cooperation with the Port

The practice of solving the problems of interaction between enterprises in a transport hub with the help of information technologies will make it possible to achieve positive dynamics of the operational performance of railway stations when interacting with ports. Specifically, in our case, the economic effect will be achieved in reducing the idle time of cars at the port station when interacting with the port and as a result of increasing the efficiency of the economic activities of the subjects of interaction.

Let us determine the cost of the idle time of cars for 2018 when the port station interacts with the port.

The average cargo train composition has 62 wagons. Base downtime rate is 3,500 Russian rubles. The value of 3500 RUB is equal to approximately 40.55 euros based on the exchange rate of 86.32 rubles per 1 euro.

The average cost of one hour downtime for a wagon freight ($C_{med}$) is:

$$C_{med} = \frac{3500}{62} = 56.45 \; \textit{russian rubles or } 0.65 \; \textit{euros}. \tag{1}$$

The local wagons idle time hours per year ($HD_L$)

$$HD_L = 32.3 \cdot 365 = 11789.5 \text{ hours of downtime per year} \tag{2}$$

The downtime cost for local freight wagons per year ($E_L$)

$$E_L = 11789.5 \cdot 56.45 = 665517.27 \; \textit{russian rubles or } 7709.88 \; \textit{euros}. \tag{3}$$

Downtime of local freight wagons without processing per year ($HD_{WP}$)

$$HD_{WP} = 0.5 \cdot 365 = 182.5 \text{ downtime hours per year}. \tag{4}$$

The cost of downtime freight wagons without processing per year ($E_{wp}$)

$$E_{wp} = 182.5 \cdot 56.45 = 10302.12 \; \textit{russian rubles or } 119.35 \; \textit{euros}. \tag{5}$$

The downtime hours of wagons with processing per year ($HD_E$)

$$HD_E = 28.4 \cdot 365 = 10366 \text{ downtime hours per year}. \tag{6}$$

The cost of downtime freight wagons with processing per year ($E_E$)

$$E_E = 10366 \cdot 56.45 = 585160.70 \; \textit{russian rubles or } 6778.97 \; \textit{euros}. \tag{7}$$

The total value of the cost of wagons downtime for the port station ($E_{COM}$)

$$E_{COM} = 7709.88 + 119.35 + 6778.97 = 14608.2 \; euros \; per \; year. \quad (8)$$

We will determine the cost of wagon downtime for 2019 when the port station interacts with the port.

The average train composition in 2019 was 62.7 wagons.

The average cost of one hour of downtime for a freight wagon (Cmed):

$$C_{med} = \frac{3500\text{ RUB}}{62.7\text{ wagons}} = 55.82 \; russian \; rubles \; or \; 0.65 \; euros. \quad (9)$$

Downtime of local wagons per year ($HD_L$)

$$HD_L = 28.7 \cdot 365 = 10475.5 \text{ hours of downtime per year.} \quad (10)$$

Cost of downtime for local freight wagons per year ($E_L$)

$$E_L = 10475.5 \cdot 55.82 = 584742.44 \; russian \; rubles \; or \; 6774.12 \; euros. \quad (11)$$

Downtime of local freight wagons without processing per year ($HD_{WP}$)

$$HD_{wp} = 1.1 \cdot 365 = 401.5 \text{ hours of downtime per year.} \quad (12)$$

Cost of downtime freight wagons without processing per year ($E_{wp}$)

$$E_{wp} = 1.1 \cdot 365 = 401.5 \text{ hours of downtime per year.} \quad (13)$$

Downtime hours of wagons with processing per year ($HD_E$)

$$HD_E = 24.5 \cdot 365 = 8942.5 \text{ hours of downtime per year.} \quad (14)$$

The downtime cost for the freight wagons with processing per year ($E_E$)

$$E_E = 8942.5 \cdot 55.82 = 499170.35 \; russian \; rubles \; or \; 5782.79 \; euros. \quad (15)$$

The total value of the cost of wagon downtime for the Zarechnaya station:

$$584742.44 + 22411.73 + 49170.35 = 1106324.48 \; russian \; rubles \; or \; 12816.55 \; euros. \quad (16)$$

The next calculation covers the downtime cost for 2020 when the port station interacts with the port.

The average train composition was 63.2 wagons.

The one hour downtime average for a freight wagon ($C_{med}$):

$$C_{med} = \frac{3500\text{ RUB}}{63.2\text{ carriages}} = 55.37 \; russian \; rubles \; or \; 0.64 \; euros. \quad (17)$$

Downtime of local wagons per year ($HD_L$)

$$HD_L = 15.5 \cdot 365 = 5657.5 \text{ hours of downtime per year.} \quad (18)$$

Cost of downtime for local freight wagons per year ($E_L$)

$$E_L = 5657.5 \cdot 55.37 = 313255.77 \textit{ russian rubles or } 3629.01 \textit{ euros.} \quad (19)$$

There is no transit without processing for this period.
Downtime hours of wagons with processing per year ($HD_E$)

$$HD_E = 13.1 \cdot 365 = 4781 \text{ hours of downtime per year.} \quad (20)$$

Cost of downtime freight wagons with processing per year ($E_E$)

$$E_E = 4781 \cdot 55.37 = 264723.97 \textit{ russian rubles or } 3066.77 \textit{ euros.} \quad (21)$$

The total value of the cost of wagon downtime for the Zarechnaya station:

$$E_{COM} = 313255.77 + 264723.97 = 577979.74 \textit{ rubles or } 6697.78 \textit{ euros per year.} \quad (22)$$

Below presented the calculation data in the form of a diagram in Fig. 4.

According to the calculations presented, there is a positive trend in wagon downtime costs reduction at the port station when interacting with the port. The comparison of 2019 and 2018 values has shown that the wagon downtime cost decreased by 154655 rubles (€1 791.65) and the comparison with 2020 numbers indicates further decrease in vehicle downtime cost – 528 345 rubles (€6120.77).

This positive trend is observed due to the introduction in 2019 at the port station of the DILS system, which is responsible for the plan for supplying trains and connecting the port to the "DILS contour", which is one of the elements of this system. Consequently, the introduction of updated information technologies is associated with the need to organize a single information field that guarantees the relevance, integrity and consistency of information that will be used to make the necessary management decisions by all parties participating in the transportation process, taking into account that each of the participants has its own automated system. This interaction of railway and water transport on the basis of the presented information system will make it possible to increase the profitability of both the port station and the port, saving the port station from idle time of loaded trains, while ships free of cargo are standing in the port.

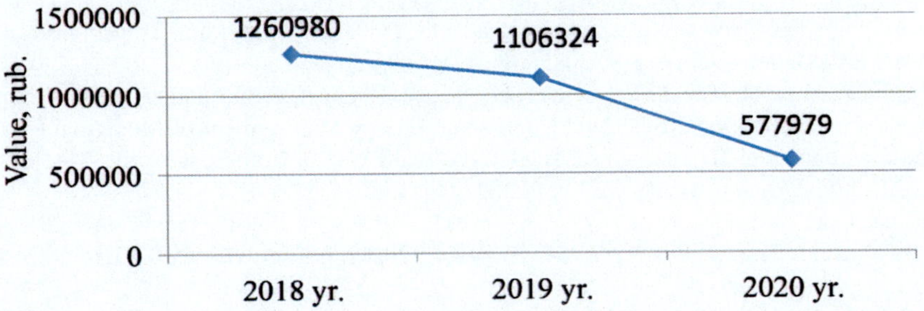


**Fig. 4.** The dynamics of wagon downtime cost reduction due to the implementation of the DILS system at the port station in cooperation with the port (commissioned in 2019)

## 4 Conclusion

In the due course of the study, the outcomes were obtained that corresponded to research and professional objectives. The developed methodology for determining the effectiveness of information integration for transport services in ports and port stations based on the organization of a single information field, including a sequence of activities, such as: - analysis of the dynamics of unloading and loading during the interaction of port stations and ports during the implementation of the Road Information and Logistics System (DILS) on the North Caucasian railway; - proposed a methodology for assessing the effectiveness of information interaction between port stations and ports based on the introduction of the "Kontur DILS" system; - approbation and economic justification of measures from the introduction of information systems at the port station when interacting with the port.

The calculations were carried out based on the method presented in this study. According to the calculations, there is a positive trend towards reducing the wagons downtime cost at the port station when interacting with the port. The 2018 and 2019 comparison indicates the cost has decreased by € 1,791.65, and 2019 period compared to 2020 shows further reduction expected by € 6,120.77.

In the process of the methodology development to optimize the interaction of all the parties involved in the transportation related processes of a transport hub based on information integration for transport services in ports and port stations, the value of the economic effect obtained as a result of reducing the idle time of wagons was taken into account. The result made it possible to determine the priority of the information system used in transport hubs.

## References

1. Aulin, V., et al.: Realization of the Logistic Approach in the International Cargo Delivery System. Communications - Scientific letters of the University of Zilina **21** (2), 3–12 (2019). https://doi.org/10.26552/com.C.2019.2.3-12.
2. Chislov, O., Bogachev, T., Kravets, A., Bogachev, V., Zadorozhniy, V., Egorova, I.: Time parameters optimization of the export grain traffic in the port railway transport technology system. In: Sierpiński, G. (ed.) TSTP 2019. AISC, vol. 1091, pp. 126–137. Springer, Cham (2020). https://doi.org/10.1007/978-3-030-35543-2_11
3. Chislov, O., Bogachev, V., Zadorozhniy, V., Bogachev, T.: Economic-geographical method delimiting wagon flows in the region considered: model and algorithm. Transp. prob. **13**(2), 39–48 (2018). https://doi.org/10.20858/tp.2018.13.2.4.
4. Vasilenko. M., et al.: The digital technologies in quality and efficiency management of transport service. In: International Scientific Conference Energy Management of Municipal Facilities and Environmental Technologies EMMFT 2020 E3S Web of Conferences. vol. 244, p. 11046 (2021). https://doi.org/10.1051/e3sconf/202124411046
5. Nechaev, A., Skorobogatova, Y., Nechaeva, M.: Toolkit for the transportation and logistics infrastructure. Transp. Res. Proc. **54**, 637–644. (2021). https://doi.org/10.1016/j.trpro.2021.02.116.
6. Kolesnikov, M.V., Lyabakh, N.N., Mamaev, E.A., Bakalov, M.V.: Efficient and secure logistics transportation system. IOP Conf. Series: Mater. Sci. Eng. **918**, 012031 (2020). https://doi.org/10.1088/1757-899X/918/1/012031

7. Safronova, A., Kurenkov, P., Egorov, A., Sokolova, A., Pilipchuk, N.: Management of the investment project structure in the field of manufacturing innovative building materials. MATEC Web Conf. **251**, 05004 (2018). https://doi.org/10.1051/matecconf/201825105004.
8. Drozdov, N., et al.: A systematic approach to modeling the production and economic activities of transport organizations. E3S Web Conf. **244**, 11045 (2021). https://doi.org/10.1051/e3sconf/202124411045.
9. Kurbatova, A., Kurenkov, P., Safronova, A., Tanaino, I.: Selection of optimal supply schemes. IOP Conf. Ser.: Earth Environ. Sci. **403**, 012231 (2019). https://doi.org/10.1088/1755-1315/403/1/012231.

# The Threats Related to Parking Electric Vehicle in Underground Car Parks

Małgorzata Krol[1](✉) and Aleksander Krol[2]

[1] Faculty of Energy and Environmental Engineering, Silesian University of Technology, Gliwice, Poland
malgorzata.krol@polsl.pl

[2] Faculty of Transport and Aviation Engineering, Silesian University of Technology, Katowice, Poland
aleksander.krol@polsl.pl

**Abstract.** Electric vehicles are increasingly appearing in underground garages. They pose a significant risk when a fire breaks out. Risks to the building structure and to the people safety increase. It is very difficult to extinguish an electric vehicle fire, and in addition, large amounts of hydrogen fluoride are released when the lithium-ion battery is burning. Substances released during an electric vehicle fire pose a threat to rescue teams, contaminate garage surfaces and contaminate the extinguishing water. The article presents numerical analyzes of the spread of hydrogen fluoride and temperature distribution during a fire of a small electric vehicle in a ten-car underground garage. The calculations used the Fire Dynamics Simulator program, which is very often applied to analyze the conditions during a fire in a confined space. Calculations show that even a fire of a small lithium-ion battery produces an amount of hydrogen fluoride that is hazardous to the health of people which could be near the fire.

**Keywords:** Underground garage · CFD · Electric vehicle · Hydrogen fluoride

## 1 Introduction

The electric vehicle market is a rapidly growing area, especially in Western European countries. There are relatively few electric cars in Poland. However, the dynamics of the increase of the number of purchased electric vehicles shows that these vehicles will appear more and more often on our roads and will be parked in underground garages. Most of the underground garages that are currently put into use have charging points for electric vehicles. The question arises whether systems installed in underground garages are prepared for parking electric vehicles. Can electric vehicles in underground facilities pose an additional risk due to the lithium-ion battery?

There is still little knowledge about the dangers of using lithium-ion batteries [1]. The need to formulate guidelines for the use of such batteries is indicated. In the aim to decrease the thermal hazards of lithium-ion batteries as much as possible, certain management or countermeasures must be implemented [2].


G. Sierpiński (Ed.): TSTP 2021, LNNS 352, pp. 72–81, 2022.
https://doi.org/10.1007/978-3-030-91156-0_6

Researches conducted so far often focused on the dangers directly related to damage to the lithium-ion battery. A risk assessment was conducted for hydrofluoric acid (HF) and lithium hydroxide (LiOH) which potential might leak from lithium-ion batteries [3] during a thermal abuse. The authors state that when a lithium-ion battery explodes in a limited space, HF emissions can reach 81.8 mg/m$^3$, in the worst-case scenario.

Larsson notes that a fire of lithium-ion batteries may generate the intense heat which can be a serious threat in many situations. But the risks associated with gas and smoke emissions from malfunctioning lithium-ion batteries may in some circumstances be a larger threat, especially in confined environments [4]. He also points out that lithium-ion batteries installed in vehicles are exposed to vibrations, moisture and possible impacts. This can create a risk related to HF emissions from batteries subjected to various influences [5]. HF emissions from car batteries are also a major risk, according to Sturk. He points out that the amount of HF emitted depends on the state of charge of the battery [6]. Some researchers are also deliberating which type of battery will be best for electric vehicles [7]. RISE (Research Institutes of Sweden) prepared report providing background information regarding lithium-ion batteries and battery pack mounting in vehicles. Fire hazards are identified and means for preventing and controlling them are presented. The possibility of fixed fire suppression systems and detection systems in electric vehicles is discussed [8].

Researches and analysis regarding the safety of electric vehicles are being carried out in some countries [9, 10] to fulfill the regulations gap. The governments have already started to define their regulatory framework for EV, mostly in order to ensure the safety of use such vehicles [11]. However, the most important issue concerns the parking electric vehicles in underground garages and the process of charging the electric vehicles in these garages. The limited space in the garage means that a lithium-ion battery fire can be a huge threat to the building structure, occupants or emergency services [12–15]. An electric vehicle fire in an underground garage can create a serious hazard due to the difficulties in extinguishing a fire of a lithium ion battery and the kind of substances released during such fires. Research confirmed that large amounts of hydrogen fluoride were released, much more than in a gasoline vehicle fire. Hydro fluoride will be a major threat to rescue teams. During the fire-fighting action, it will contaminate the fire-fighting water, which is then discharged into the sewage system. After the fire is extinguished, hydrogen fluoride may remain on the surface of the partitions and it may be necessary to decontaminate the underground garage walls [8, 9].

The paper presents numerical analyzes of an electric vehicle fire in a small underground garage. The Fire Dynamics Simulator (FDS) was used in the calculations. The research focused on the temperature distribution and the concentration of hydrogen fluoride in the garage space during the fire development.

## 2 A Numerical Model of a Small Underground Car Park

Numerical calculations were made with the use of the FDS program [16]. Modeling of fire development is a very complex task. It involves numerical solving of flows, combustion and heat transfer. The well recognized FDS (Fire Dynamic Simulator) software was applied in the presented work. It is a software package developed by National Institute of Standards and Technology (NIST, USA), dedicated for low Mach Number, buoyancy driven flows. FDS implements Finite Volumes Method (FVM), in which the whole computational domain is divided into small cells. For each such volume, balance equations are solved for all quantities describing fluid flow (mass, energy, momentum). FDS by default applies Large Eddy Simulation (LES) approach for solving turbulent fluid flows. Since a significant part of eddies is smaller than cell size, in the LES method a low-pass Favre filter is applied to the transport equations. Hence, in this way just mean values of quantities being considered are calculated for each cell, meanwhile larger eddies are solved directly [16].

The combustion process is modeled using mixture fraction model. In this approach chemical reactions of combustion are regarded as immediate in relation to flows (what is mixed is burnt). The mixture of gases contains oxygen, nitrogen, fuel components and final (like water vapor and carbon dioxide) and intermediate products of reactions. Since the radiative heat transport plays an important role in fires modeling and fire gases commonly contain large amounts of optically dense products of combustion, the full equation of radiation transport must be solved. It covers emission, absorption, scattering and refraction of thermal radiation. As soot, which optical properties are not particularly sensitive to wavelength is the dominant source and sink of thermal radiation, a simplification called grey gas model can be applied for more efficient calculations.

The underground garage considered in the analyzes has an area of 384 $m^2$ (Fig. 1). According to the regulations, the garage might be equipped with natural ventilation and no fire ventilation is required. However the exhausting ventilation was assumed which turns on when a fire breaks out. It consists of an exhaust vent measuring 0.8 x 0.4 m, located in the back wall of the garage. Fresh air is to flow into the garage through the entrance gate, opened automatically when a fire breaks out. There are 10 parking places in the garage.

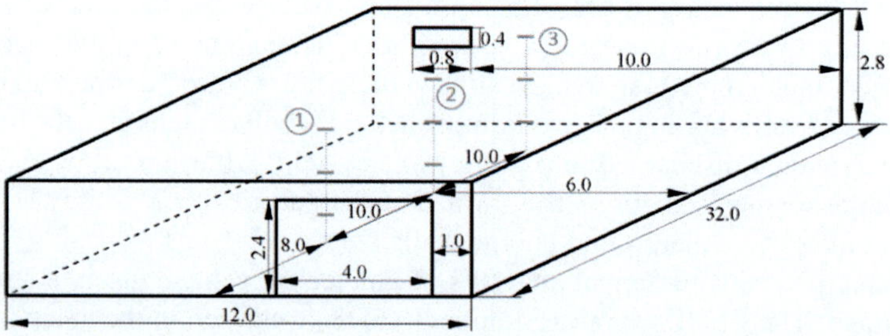


**Fig. 1.** Underground garage used for numerical analyzes. Drawing not to scale. Measurements in meters.

On the basis of data showed in Fig. 1, a numerical model of the garage was built in the FDS program (Fig. 2). Two models were built that differed in the number of elements. In the first model the computational grid contained 1677312 elements and the dimensions of a single element were 0.1 × 0.1 × 0.1 m. In the second one the computational grid contained 209664 elements and the dimensions of a single element were 0.2 × 0.2 × 0.2 m. This allowed for confirming the results independence on the density of the numerical mesh.

The grid was extended beyond the garage area near the wall with a gate so the out flowing smoke could be observe as the fire unfolded within the garage. It has been assumed that there are three vehicles parked in the underground garage. One of these vehicles is an electric one. A fire breaks out due to a short circuit in a lithium-ion battery of the electric vehicle. It is assumed that the electric car is small and is powered by a small lithium-ion battery. The battery parameters are as follows: nominal voltage 330 V, capacity 50 Ah and stored energy 16.5 kWh. The battery is in 100% charged. Based on Lecocq's [10] research, the heat release rate and the emission of hydrogen fluoride were assumed (Fig. 3).

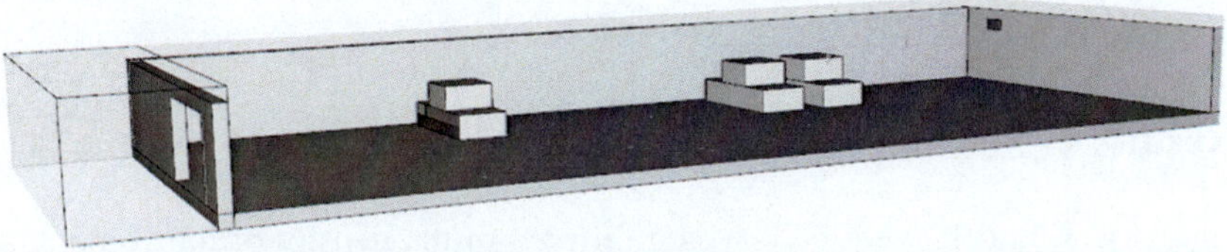

**Fig. 2.** Numerical model of the underground garage built in the FDS program.

As was mentioned earlier, an underground garage of the same size as in the analyzed example does not need to be equipped in a mechanical ventilation. According to the regulations, it is enough to equip such a garage with a natural ventilation. The operation of natural ventilation is based on the flow of air through the garage caused by the difference in the density of internal and external air and the effect of the external wind. In such situation modeling a natural ventilation system would require a significant expansion of the computational domain. This would result in the appearance of a natural airflow on its own as an effect of the set boundary conditions. However, this would significantly extend the time of calculations and would require consideration of many variable parameters of the external climate (different values of temperatures and wind direction and velocity). Therefore, a simplification was introduced to the model by assuming the exhaust mechanical ventilation in the garage.

The capacity of such mechanical ventilation was adjusted to mimic the operation of the natural system. Garage air exchange was assumed to be 1.5/h changes, hence an outflow of 0.5 $m^3$/s was assumed through the exhaust vent in the rear part of the garage, the compensation air flowed into the garage through the entrance gate. The introduced simplification describes a system that could be used in such a garage and fulfils the regulations requirements. It would provide greater safety in the event of a fire because it would force the removal of a specific volume of fire smoke from the garage space regardless of external weather conditions.

In order to observe the temperature distribution and the concentration of hydrogen fluoride, virtual sensors were introduced in the numerical model. They were placed on three stands. Each stand has three points for measuring the temperature and concentration of hydrogen fluoride at a height of 0.8, 1.8 and 2.8 m. Locations of the stands are shown in Fig. 1.

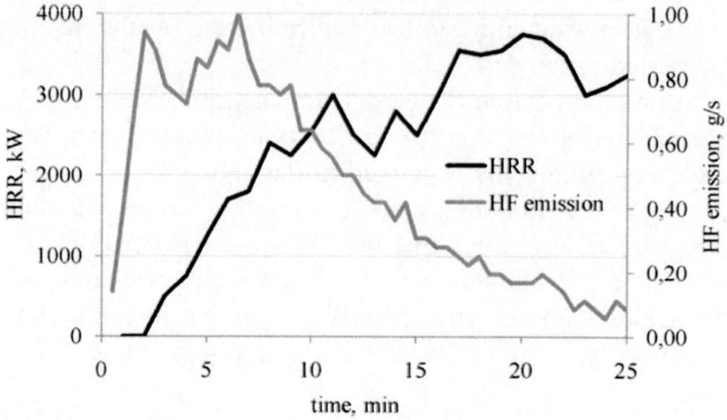


**Fig. 3.** The heat release rate of the fire and the emission of hydrogen fluoride during the fire development [based on 10].

# 3 Results

Two numerical models were built which differed in the density of the computational grid. This allowed for performing the analysis of the mesh sensitivity. Figure 4 shows the concentrations of hydrogen fluoride at three different heights for the stand number 1, which was closest to the burning vehicle. The results for the model with a coarse mesh and for the model with a dense mesh are presented.

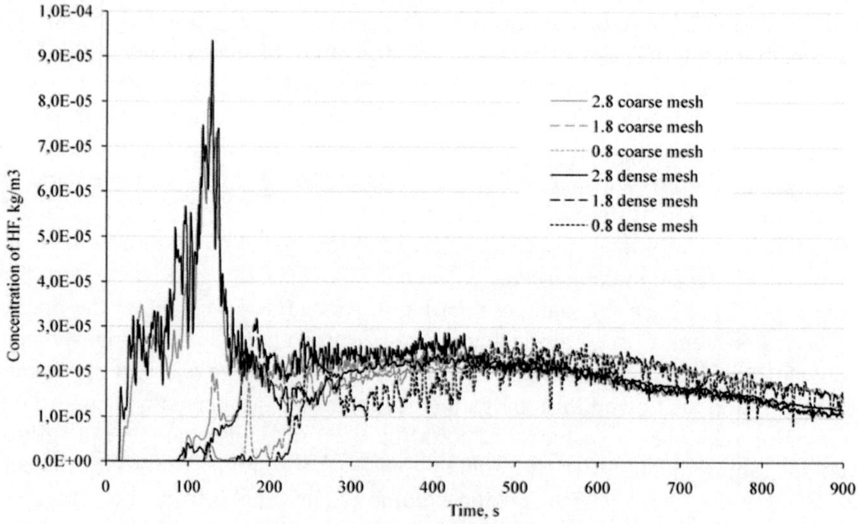


**Fig. 4.** Hydrogen fluoride concentration at three different heights for two different mesh densities.

Concentration values obtained for the coarse mesh and for the dense mesh are very consistent. For the same heights of the measurement point, the curves have a similar course. Due to the greater accuracy of the calculations, the results for a denser mesh are taken into account in the following considerations.

The dangers of an electric vehicle fire in an underground garage are related to the difficulties in extinguishing such a fire and releasing significant amounts hydrogen fluoride and other poisonous chemicals. Many researchers point out that the amount of hydrogen fluoride released may be up to two times greater than during a fire of a vehicle with petrol engine [10].

Polish regulations define the maximum permissible concentrations of hydrogen fluoride depending on the exposure time, the values are given in Table 1.

**Table 1.** Maximum permissible concentrations of HF [17].

| Exposure time | Concentration, $mg/m^3$ |
|---|---|
| Maximum permissible concentration during a work shift | 0,5 |
| Instantaneous maximum permissible concentration, not longer than 15 min | 2,0 |
| Lethal concentration | 40 |

The performed numerical analyzes allowed to observe the spread of the HF from a burning small lithium-ion battery. The following figures show the concentration of HF at three points in the garage according to the stands position (Fig. 1). Each plot shows time changes in the concentration of HF at a certain height. Additionally, the level of the instantaneous maximum permissible concentration and the level of the lethal concentration are marked at each graph.

The concentration of HF just under the garage ceiling is high. For several dozen seconds it exceeds the lethal concentration, especially near a burning electric vehicle (Fig. 5). The shape of the HF concentration curve at a height of 2.8 m is consistent with the literature [10]. At about 140 s the concentration drops below the lethal concentration and remains below this value until the end of the simulation. The instantaneous maximum permissible concentration is exceeded for the entire duration of the simulation.

The HF concentration at the other two heights remains below the lethal concentration (Fig. 6 and Fig. 7). It should be noted, however, that the instantaneous maximum permissible concentration is exceeded at all points where the HF concentration is measured in the garage during the electric vehicle fire.

The presented analyzes assume a fire of a small lithium-ion battery. This was due to the limited availability of data describing HF emissions from larger lithium-ion batteries. It can be assumed that the HF emission from higher capacity batteries will be respectively higher. The type of ventilation system operating in the underground garage will also have a direct impact on the level of HF concentration in the garage space. The presented analyzes assume a constant air exchange in the garage at 1.5/h. In fact, in such a small underground garage, there is no requirement to install mechanical ventilation. This means

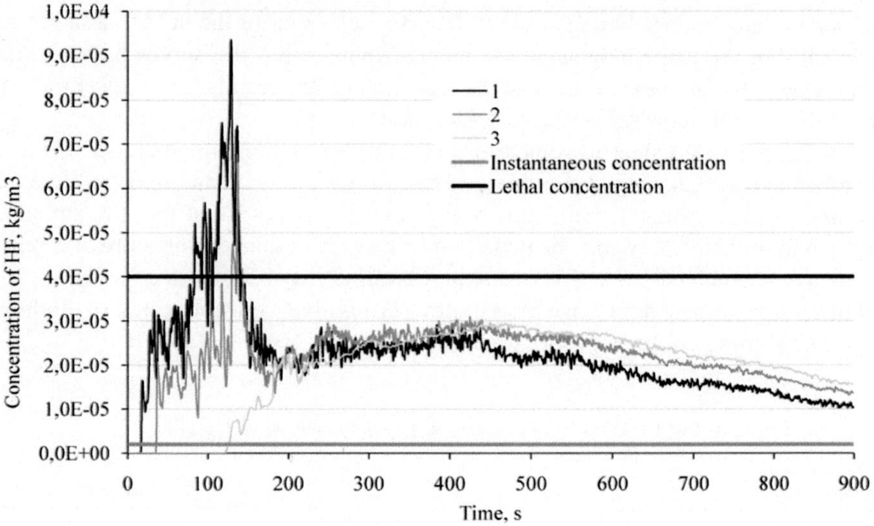


**Fig. 5.** Concentration of HF at a height of 2.8 at three points in the garage. Values of lethal concentration and instantaneous maximum permissible concentration are marked also.

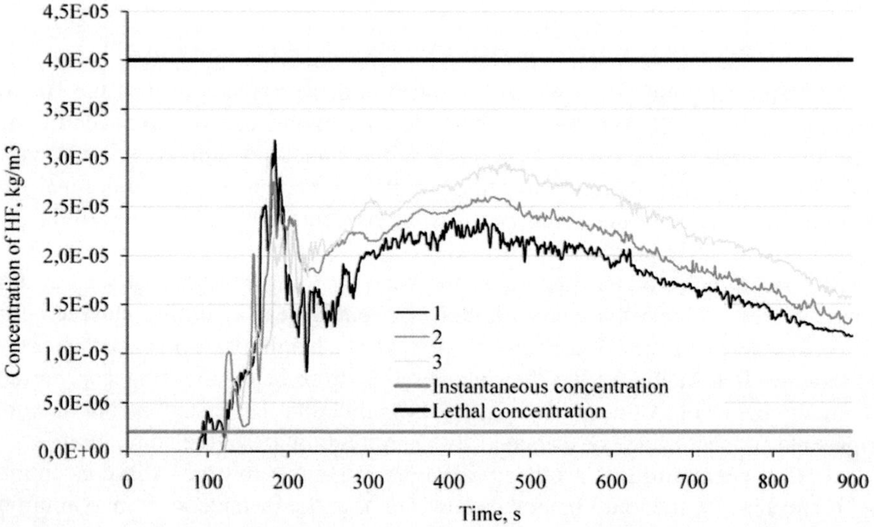


**Fig. 6.** Concentration of HF at a height of 1.8 at three points in the garage. Values of lethal concentration and instantaneous maximum permissible concentration are marked also.

that the air exchange depends only on the operation of natural ventilation, for which it is very difficult to determine the overall efficiency. This can make the conditions in the garage with only natural ventilation much worse than those shown in the simulations.

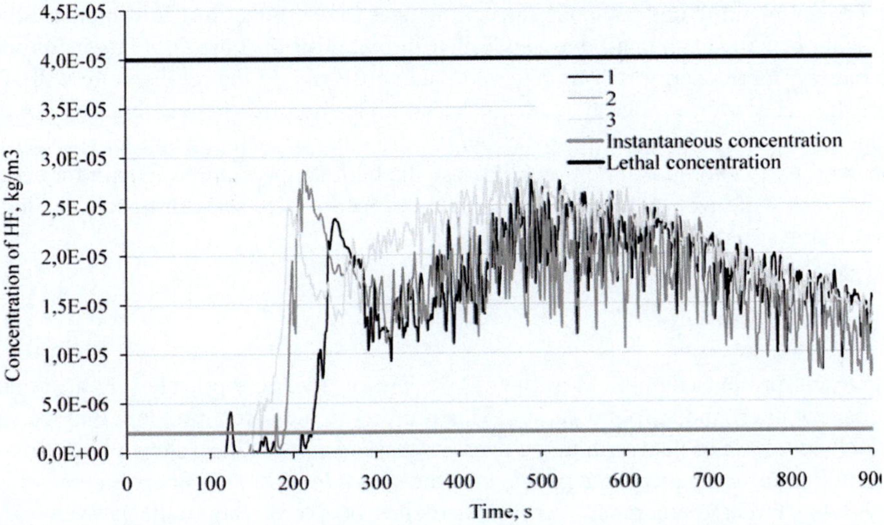


**Fig. 7.** Concentration of HF at a height of 0.8 at three points in the garage. Values of lethal concentration and instantaneous maximum permissible concentration are marked also.

The temperature distribution at the same points as the HF concentration was also monitored. Figure 8 shows the temperature values under the garage ceiling at three points, according to the position of the stands (Fig. 1).

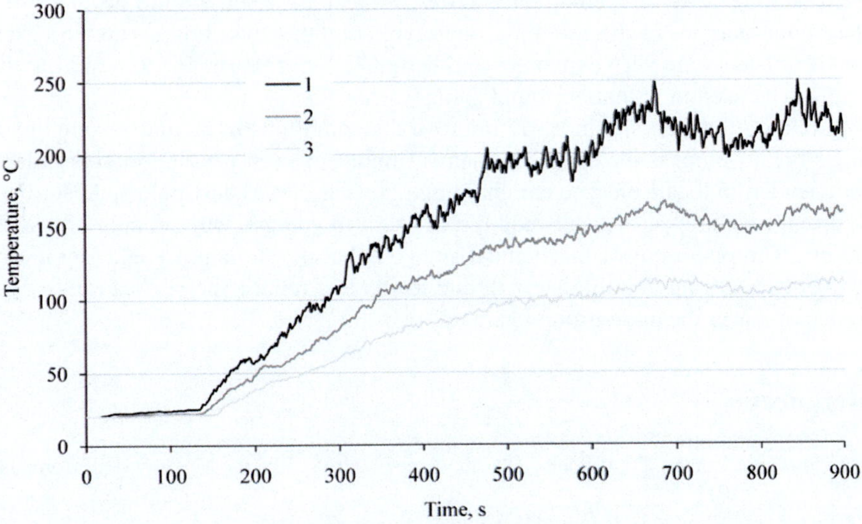


**Fig. 8.** Temperature at three points in the garage at a height of 2.8 m.

The temperature under the garage ceiling near the burning electric vehicle reaches the value of 250 °C. It can be assumed that in the event of a fire of a larger lithium-ion battery, more energy will be released, and consequently the temperature will also be higher. However, in a case of an electric vehicle fire, the greatest threat is not the temperature of the surrounding space, but the difficulty in extinguishing of such a fire. The inability to extinguish the fire will cause the high temperature to remain for a long time, which will favor the further development of the fire. This can ultimately endanger the building structure.

## 4 Conclusions

The article presents numerical analyzes of a fire of a vehicle powered by a lithium-ion battery in an underground garage. The aim was to indicate a problem that we will undoubtedly have to deal with in the nearest future. Analyzes have shown that a fire of a small lithium-ion battery in a garage generates such level of HF concentration, which may pose a threat to emergency services and other people staying in the garage even in the initial stage of the fire.

HF from a battery fire will be deposited on the garage walls. Its poisonous effect will last even after the fire is extinguished. The decontamination of the garage space is extremely important in this situation.

The discharge of extinguishing water from extinguishing an electric vehicle directly to municipal sewage is also a significant threat to the environment. Such water is highly contaminated and should undergo a special treatment process.

The difficulties in extinguishing a electric vehicle fire are well known. However, these difficulties take on a completely different meaning when the fire develops in the underground garage. In the literature, there are solutions based on sprinkler systems. Sprinkler systems are very expensive and there is a lot of resistance to requiring these systems to be used in all underground garages.

A solution that can significantly improve the conditions in an underground garage where a fire develops is the use of mechanical smoke exhaust ventilation. It is generally recommended to locate electric car charging points at the exhaust points. According to the current regulations, it is not required to use it in garages with an area of less than 1500 $m^2$. The operation of mechanical smoke exhaust ventilation could significantly improve the working conditions of rescue teams and reduce the risk caused by large amounts of HF in the underground garage.

## References

1. Reddy, T.B., Linden, D.: Linden's Handbook of Batteries, 4th edn. McGraw-Hill Companies, New York (2011)
2. Ouyang, D., Chen, M., Huang, Q., Weng, J., Wang, Z., Wang, J.: A review on the thermal hazards of the lithium-ion battery and the corresponding countermeasures. Appl. Sci. **9**, 2483 (2019). https://doi.org/10.3390/app9122483
3. Park, Y.J., Kim, M.K., Kim, H.S., Lee, B.M.: Risk assessment of lithium-ion battery explosion: chemical leakages. J. Toxicol. Environ. Health **21**(6), 370–381 (2018)

4. Larsson, F., Andersson, P., Blomqvist, P., Mellander, B.-E.: Toxic fluoride gas emissions from lithium-ion battery fires. Sci. Rep. **7**, 10018 (2018). https://doi.org/10.1038/s41598-017-09784-z
5. Larsson, F., Andersson, P., Mellander, B.-E.: Lithium-ion battery aspects on fires in electrified vehicles on the basis of experimental abuse tests. Batteries **2**(2), 9 (2016). https://doi.org/10.3390/batteries2020009
6. Sturk, D., Hoffmann, L., Tidblad, A.: Fire tests on e-vehicle battery cells and packs. Traffic Inj. Prev. **16**, 159–164 (2015)
7. Iclodean, C., Varga, B., Burnete, N., Cimerdean, D., Jurchiş, B.: Comparison of different battery types for electric vehicles. In: IOP Conference Series: Materials Science and Engineering, vol. 252, p. 012058 (2017). https://doi.org/10.1088/1757-899X/252/1/012058
8. Brandt A.W., Glansberg K.: Charging of electric cars in parking garages, RISE-report (2020)
9. Valisalo, T.: Firefighting in case of Li-Ion battery fire in underground conditions: literature study, RESEARCH REPORT VTT-R-00066-19 (2019)
10. Lecocq, A., Bertana, M., Truchot, B., Marlair, G.: Comparison of the fire consequences of an electric vehicle and an internal combustion engine vehicle. In: International Conference on Fires In Vehicles - FIVE 2012, pp.183–194, Chicago, United States (2012)
11. Global Technical Regulation on the Electric Vehicle Safety. ECE/TRANS/180/Add.20. May (2018)
12. Boehmer H., Klassen M., Olenick S.: Modern Vehicle Hazards in Parking Structures and Vehicle Carriers, NFPA (2020)
13. Boe A.G., Reitan N.K.: Fire risks of electrical vehicles in underground car parks. In: Fourth International Conference on Fire in Vehicles, vol. 243, 5–6 October, Baltimore, USA (2016)
14. Risk minimization of electric vehicle fires in underground traffic infrastructures, Federal Department of the Environment, Transport, Energy and Communications DETEC, Research project AGT 2018/006 petitioned by the Tunnel Research Working Group, August (2020)
15. Bisschop, R., Willstrand, O., Amon, F., Rosengren, M.: Fire Safety of Lithium-Ion Batteries in Road Vehicles, RISE Report (2019)
16. McGrattan, K., McDermott, R., Vanella M., Hostikka, S., Floyd, J.: Fire Dynamics Simulator. Users Guide. NIST Publ.1019, Gaithersburg, MD (2020). https://doi.org/10.6028/NIST.SP.1019
17. Regulation of the minister of labor and social policy 1 of 6 June 2014 on the maximum allowable concentrations and intensities of factors harmful to health in the work environment. J. Laws, 1348 (2017)

# Intelligent and Integrated Transport Systems

# Modeling Multitasking Onboard of Privately-Used Autonomous Vehicle and Public Transport

Jamil Hamadneh(✉) and Domokos Esztergár-Kiss

Department of Transport Technology and Economics, Faculty of Transportation Engineering and Vehicle Engineering, Budapest University of Technology and Economics (BME), Budapest 1111, Hungary
jamil.hamadneh@kjk.bme.hu, esztergar@mail.bme.hu

**Abstract.** Autonomous Vehicles (AVs) will impact the current travel behavior of people because the characteristics of AV are different than conventional cars. In AV, the machine takes the role of driving rather than a human driver. This study adds a new contribution to the literature by studying privately-used autonomous vehicle (PAV) and public transport (PT) considering multitasking availability. The focus of this research is on the influences of multitasking options availability, travel time, and travel cost on the travelers' behavior. In this research, six options of multitasking are defined, such as reading, writing, talking, using social media, eating-drinking, and doing other things. A stated choice experiment is designed, where people are asked to choose either PAV or PT based on their preferences and the characteristics of each transport mode. A sample size of 4366 observations is collected in Budapest, Hungary at the beginning of 2020. A discrete choice modeling approach is applied through the Conditional Logit model to develop a transport choice model and find the impact of attribute's changes on the choice of travelers. In this research, the main daily trip inside urban areas is only studied. The results show that only writing onboard adds a negative utility in choosing a transport mode, while others add a positive utility. Moreover, gender and car ownership variables impact the selection of a transport mode. In conclusion, people are more likely to choose PAV over PT inside urban areas because the margins show a larger modal share of PAV than PT at the same trip time and trip cost.

**Keywords:** Multitasking · Autonomous vehicle · Discrete choice model · Travel behavior · Value of travel time

## 1 Introduction

Travelers are rational in their decisions, as stated in the random utility theory [1]. Travelers choose a transport mode to maximize their utility by choosing a suitable transport mode, route, companion, carried tools, departure time, and arrival time to have a pleasant journey. A transport mode is chosen based on travelers' preferences. For example, crowding in public transport, waiting time at stop station, and comfort in-vehicle are


G. Sierpiński (Ed.): TSTP 2021, LNNS 352, pp. 85–100, 2022.
https://doi.org/10.1007/978-3-030-91156-0_7

factors that can be controlled by the traveler when he/she chooses a transport mode (i.e., factors are compensated by money) [2]. The value of travel time (VOT) reflects the importance of traveling by a transport mode based on traveler preferences [3]. Having pleasant in-vehicle time affects the amount of VOT, for example, the pleasant journey has a lower VOT than an unpleasant one. Onboard activities decrease the VOT, while the amount of VOT is affected by other factors, such as trip purpose, and the characteristics of travelers (i.e., the high-income group has high VOT than the low-income group). Moreover, conducting onboard activities is associated with the carried tools onboard by travelers and the tools provided by the operators, for example, information and communication technology (ICT) [2, 4–6]. Researchers evaluate the travel time based on three parts, walking time, the in-vehicle time, and the waiting time [7]. Researchers find that waiting time has the highest VOT compared to others [7, 8]. In the autonomous vehicles (AVs) era, the AVs will minimize the walking time and provide a better environment onboard; AV is considered door-to-door services. Moreover, PAV has unique characteristics that are considered superior compared to conventional cars, for example, AV provides machine drivers, higher safety, and remove the need for driving license to use [9].

Multitasking in conventional transport modes is studied by Varghese and Jana [4], and Ettema and Verschuren [10]. They find changes in the VOT due to multitasking. Malokin, Circella [11] show that traveler changes his/her transport mode based on the availability of multitasking and carried technological tools. The attitudes of travelers onboard are studied by Rhee, Kim [12], and Mokhtarian, Papon [13], where onboard activities are presented. Etzioni, Hamadneh [14] examine the acceptance of people across seven countries to the AV. Polydoropoulou, Tsouros [15] study the willingness of people to share autonomous vehicles across seven countries. Further research is required about the AV, such as travel-based multitasking and the impact of the onboard activities on the travel utility. The travel-based multitasking method means the traveler chooses a transport mode based on the possibility of doing certain onboard activities [16]. Banerjee and Kanafani [17] say that onboard activities influence the VOT significantly of workers. Pudāne, Molin [18] demonstrate the obtained benefits from the introduction of AV, where AV is considered a good environment for conducting onboard activities and saves time for conducting other activities. Keseru and Macharis [19] show that multitasking is presented differently in several studies, and they point out the need for a unified or new definition for multitasking onboard of transport modes that can be used for later studies. Pawlak [20] collects previous papers on ICT and multitasking, he highlights the importance of finding a consensus definition of multitasking. This paper focuses on the in-vehicle time of privately-used autonomous vehicles (PAVs) (i.e., driverless car owned by manufacturers or operators, where a passenger uses it privately during his/her journey) and public transport (PT), where a traveler chooses the suitable outcome (i.e., PAV or PT) based on his preferences and the possibility of multitasking.

This study adds a new contribution to the literature, where PT and PAV are modeled based on multitasking during traveling to the main destination. Previous studies focus primarily on the analysis of conventional transport modes, while this research studies the influence of certain types of multitasking on the travel behavior of people in case of PAVs and PT. The contributions of this paper are (1) to develop transport mode choice models of PAV and PT considering multitasking options, travel time, and travel cost, and (2) to analyze the effect of each onboard activity on the transport mode choice.

## 2 Literature Review

The technology advancement is growing fast might lead to having fully automated vehicles in the near soon [21]. Having autonomous vehicles (AVs) in the market will lead to a gradual change in the current behavior of people because AVs have different properties than conventional cars [21, 22]. The influence of introducing AVs to the market depends on whether people accept this new technology or not based on their preferences. Generally, travelers search for an affordable transport mode that can ease their movements and make their traveling joyful or pleasant. Zhong, Li [23] shows that travel time, travel cost, and the possibility of multitasking are factors that impact traveler's choice. For example, the valuation of travel time onboard of AVs should be less than conventional cars in order to let people leave their cars and switch to AVs. This does not mean that AVs will be only an opponent for conventional cars, but also other transport modes, such as public transport modes. Having less VOT is attained through providing more convenient transport mode, reduction in the travel time, reduction in the travel cost, and other factors. Travelers can pay money to overcome the unwanted travel time in order to maximize their utility [24]. A reduction in the travel time means more time is allocated activities time; travelers gain utility by decreasing the negative utility of time (i.e., traveling is considered a consumption based on the random utility theory) [25–27].

People seek for low VOT during their travel. Litman [8] shows that working on the qualitative measures is less costly than working on the infrastructure measures, for example, providing comfort travel onboard is less costly than constructing a new lane or new road. He demonstrates that the VOT is estimated by comfort, convenience, and safety aspects rather than solely the reduction of travel time. The importance of multitasking onboard is revealed in the perceived travel time [2], and it is found that travelers can travel longer distances if the travel is pleasant onboard [28]. Xumei, Qiaoxian [29] find that waiting time has the highest value of VOT, while leisure trip has lower VOT than work trips. Athira, Muneera [30] say that VOT is increased when trip time and income are increased. The advancement of technology provides tools that can participate in improving the perceived travel time in-vehicle. Besides, the AV is an advanced technology that might provide a suitable environment and extra tools onboard. AV enhances the mobility of people where all users are passengers, no driving license is required, and the possibility of multitasking is increased [27, 31, 32].

Ian Wallis Associates Ltd [33] reports several findings such that the VOT of the passenger is third-fourths of the VOT of the driver. Moreover, an increment in the VOT is noticed when drivers drive long distances and with the presence of more than one passenger. The VOT is affected by trip purpose, for example, commuting trips have higher VOT than other trip purposes [33]. However, in the AVs, no drivers are needed since the machine drives the vehicle (i.e., drivers are transformed into passengers). Generally, the VOT of nowadays drivers will be decreased with the introduction of AVs.

The discrete choice modeling approach is used to model the preferences of people towards certain alternatives based on some factors, for example, transport modes with travel time and subjective attributes [25].

Simoni, Kockelman [34], Kolarova, Steck [35], and Bozorg and Ali [36] show that AV will have low VOT than conventional cars. Gurumurthy and Kockelman [37] and Steck, Kolarova [5] find that people are more likely to use PAV over the shared-autonomous vehicle (SAV) (i.e., other passengers might share with you the journey). Lavieri and Bhat [38] find that the companion types impact and the waiting time impact the use of SAVs. Commuters have lower VOT when they use PAV or SAV [36]. Other studies show a decrease in the VOT of travelers, when they multitask onboard where a comfortable environment is provided (i.e., no the stress and mental tension generated from driving) [22, 39]. A significant reduction of VOT is obtained in long-distance travel when people can engage more in multitasking. Singleton [22] shows that long trips are better for multitasking than short trips, where a reduction in VOT is more in long trips. Wadud and Huda [39] show that the type of onboard activities is associated with the trip purpose.

Travelers try to reduce the travel time because there is a lack of a suitable onboard environment that gives them the possibility to convert part of their travel time from unproductive to productive time. In this study, PAV and PT are studied, where onboard activities are divided into six categories. This study aims to find the impact of each multitasking option on traveler behavior. The utility of every onboard activity is estimated in this study. Moreover, the impact of PAV on the PT is discussed.

## 3 Methodology

The influence of multitasking options on transport mode choice is examined through discrete choice modeling, where the two transport modes, PAV and PT are modeled using one model of the random utility theory [40]. In this study, the Conditional Logit model is used to examine the impact of transport mode characteristics on the behavior of travelers. PAV is still not on the market; a stated preference (SP) survey is conducted which includes a stated choice experiment to collect the preferences of travelers in urban areas when they conduct their main trips. An SP includes sociodemographic, economic, trip characteristics, and discrete choice experiments, as shown in Fig. 1 The discrete choice experiment is designed that considers three attributes; travel time, travel cost, and multitasking options. Three levels of travel time, three levels of travel cost, and six options of multitasking are defined. The sociodemographic variables are age and gender. Economic variables are income, job, and car ownership. Trip variables are frequent transport mode, and the frequent trip purpose (i.e., the most frequent transport mode is used to get the most frequent trip purpose by a respondent). Multitasking is defined based on the literature and the author contribution, as follows:

- Reading: travelers can read using their books/laptops/work documents (i.e., other things than social media).
- Writing: travelers can write on paper/document/laptop.
- Talking: travelers can talk with companions or their cellphones/tablets/laptops to make calls using social media.
- Using social media: using the internet to open different types of social media. The travelers can watch, listen, and interact in silence (i.e., no calls).
- Eating-drinking: travelers can eat/drink onboard of PT and PAV.

- Doing other things: doing other things than the aforementioned five points, for instance, window gazing, thinking, and sleeping.

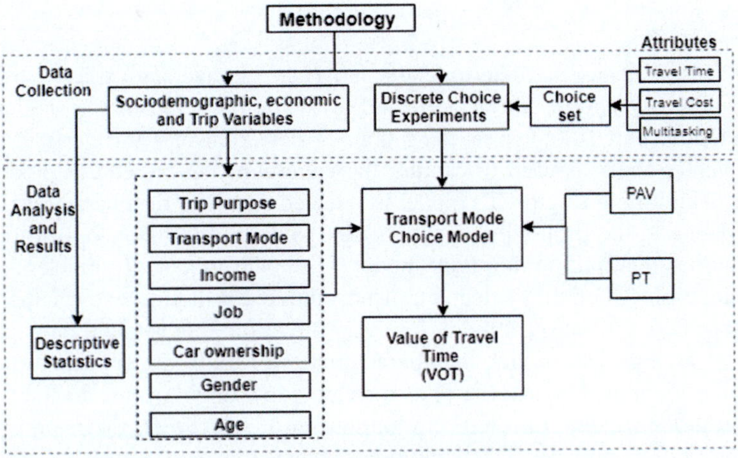


**Fig. 1.** The methodological approach

It is worth mentioning that an SP survey is conducted in Budapest, Hungary, in March/April 2020 using the LimeSurvey tool [41]. The participants are asked to record their existing conditions as well as their preferences towards PAV and PT considering the time before COVID-19, if they have been affected by the pandemic at the time of the survey.

Random utility theory is used to develop models that help in understanding the behavior of respondents. The discrete choice experiment gives an estimation for multiple alternatives rather than one alternative, and it can give an estimation to the VOT [42]. A fractional factorial design is used to design the choice experiments which includes combinations of PAV and PT with their associated attributes and levels [43]. The efficiency of the stated choice experiment method is high because it is useful in obtaining responses from participants where their preferences are associated with characteristics of transport modes they choose [40]. The analysis aims to develop mode choice models for PAV, and PT.

In the random utility theory, every traveler is considered a rational decision-maker, where each person seeks a decision to maximize the utility from doing an activity (i.e., traveling and activity utilities) [1, 44]. The probability of a traveler to choose an alternative (e.g., PAV or PT) out of certain outcomes (i.e., PAV and PT) is shown in Eq. 1:

$$Prob.(\mathrm{s}_j/\mathrm{S}) = \mathrm{Prob.}(Ut_{\mathrm{s}_j} > Ut_k)\forall k \neq \mathrm{s}, K \in S \tag{1}$$

where Prob. (sj/S) is the probability of a traveler to select transport mode (si) from the certain choice set (S), Ut (sj) stand for the utility of selecting the transport mode (sj) from the certain choice set (S), where the obtained utility is larger than the utility of other transport modes (K) in the certain choice set (S) [1]. The utility that a person gets when he/she makes daily activities includes two parts; deterministic and stochastic, as given in Eq. 2:

$$Utility_j = Det._j + \varepsilon_j \tag{2}$$

where Det. j means the deterministic part (i.e., the average utility that travelers perceive when they choose a transport mode j) while the ε means the stochastic part [1]. The random part is known as the unknown deviation of a traveler's utility from the mean value, and it is used to capture the uncertainty in the choice model [1]. In this study, Conditional Logit (CL) model is selected based on the characteristics of the collected data and the properties of this model compared to other logistic regression models, such as the multinomial logit model. In the CL model, all travelers are given different situations when they select a transport mode. Furthermore, the same travelers are asked to choose a transport mode for every situation; this is taken into account by the CL model. In the CL model, the characteristics of transport modes are taken into account rather than the characteristics of travelers [45]. Let Y (ij) means the properties of transport mode (j), and α is the vector parameter of selecting the transport mode (j), the probability that a traveler (i) will choose transport mode (j) is given in Eq. 3.

$$P_{ij} = \exp\left(Y_{ij}\alpha\right) / \sum\nolimits_{k=1}^{J} \exp(Y_{ik}) \tag{3}$$

The likelihood function is estimated using Eq. 3.

The analysis is conducted using Stata 16 [46], where the relative value of one alternative to others is obtained based on coefficients. The interpretation of the outputs is based on a reference alternative (base alternative), for example, PAV alternative estimates are referred to PT alternative [40].

## 4 Results and Discussion

The result of the SP survey is 525 respondents. The individual characteristics of respondents are summarized in Fig. 2 and Fig. 3.

Figure 2 shows the age, job, income, and gender groups. The majority of the sample is workers with age 25–54 years, and with low income. Besides, the percentage of males and females is 51.43%, and 48.57%, respectively. The high income (higher than 1100Euro) is 18.86%, and the middle income (higher than 650 Euro and equal or less than 1100Euro) is 21.33%, and the low income (equal or lower than 650 Euro) is 38.10%. Besides, 21.71% do not declare their net monthly income.

Figure 3 presents the transport mode and trip purpose of each respondent. The modal share of the sample is presented, where the largest modal share is public transport (41.71). The main trip purpose inside urban areas that participants reach is work (50.10%). Moreover, 37.71% of persons travel to education (i.e., students).

Table 1 shows the results of the model. It is shown that the sign of travel time and travel cost is negative, which complies with the random utility theory that says traveling is disutility. The VOT is 1097 Ft/hour which is calculated as the ratio of the

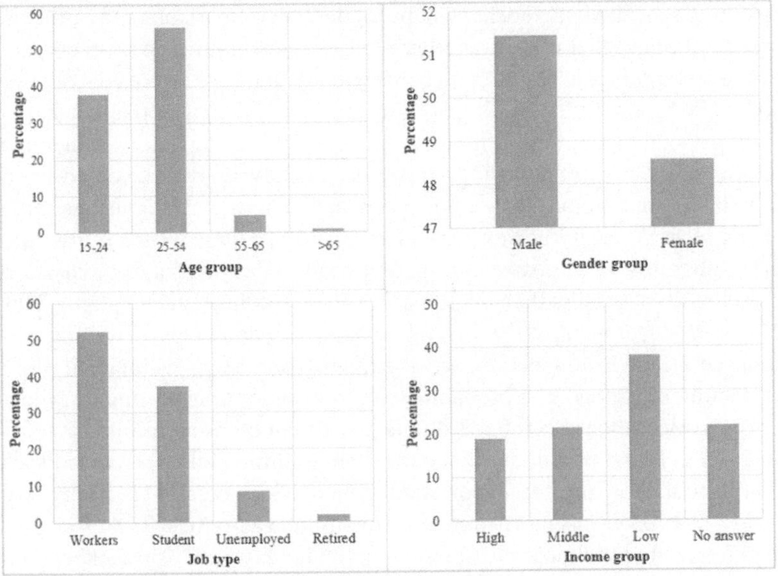


**Fig. 2.** Sociodemographic and economic variables: age, gender, job, and income

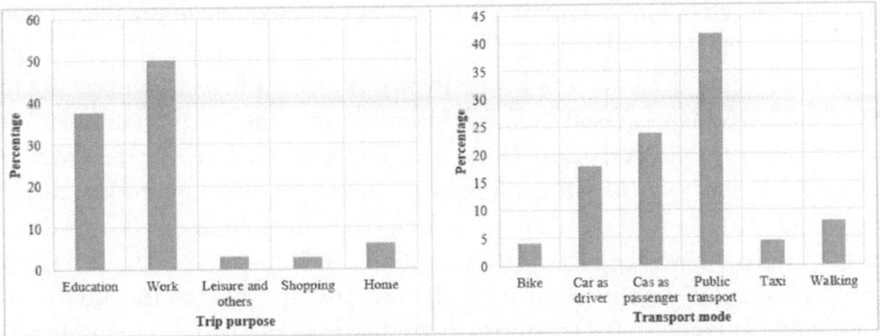


**Fig. 3.** Trip characteristics variables: transport mode and trip purpose

marginal utility of travel time (0.0311) to the marginal utility of travel cost (0.0017). Multitasking options show varied results, such as writing shows a negative impact on the transport mode choice, while other multitasking options show positive impacts with varied values. The attributes of the transport choice model are not all significant at a confidence interval of 95%. For example, travel time, travel cost, and multitasking-talking are significant at confidence interval of 95%. Multitasking-reading and multitasking-using social media are significant at a confidence interval of 90%. While multitasking-writing and multitasking-eating/drinking are insignificant. Furthermore, the variations across groups are examined, such as gender, and car ownership. The two variables are kept in the model because they show plausible significant results, at a confidence interval of 80%, and 95%, respectively. Moreover, the model is enhanced when gender and car ownership are included.

The insignificant attributes are kept in the model for comparisons and further analysis. It is worth mentioning that other sociodemographic, economic, and trip characteristics variables show insignificant results, and consequently, they are removed from the model. In the model, the base activity is doing nothing (i.e., other activities than reading, talking, writing, using social media, and eating/drinking), and it is called later as other mentioned activities. The probability of choosing a transport mode when reading activity is possible onboard is 0.236 points more than other mentioned activities. This means that reading activity adds positive utility because travelers spend productive time during their journeys. The probability of choosing a transport mode when writing activity is possible onboard is 0.0267 points less than other mentioned activities (i.e., negative). This means that writing adds negative utility, which means travelers do not enjoy writing onboard or they do not see the environment onboard as suitable for writing (vehicle vibration, and a place for document/paper). The probability of choosing a transport mode when talking activity is possible onboard is 0.0223 points more than other mentioned activities. The talking activity is found productive since travelers minimize their perceived travel time. The probability of choosing a transport mode when using social media activity is possible onboard is 0.1903 points more than other mentioned activities. Therefore, the benefit of the advancement of technology is demonstrated here, where using social media adds the second largest positive utility to the traveler utility. While reading activity adds the largest utility to travel utility. It is worth mentioning that using social media is not free while reading is free. The probability of choosing a transport mode when eating/drinking activity is possible onboard is 0.0049 points more than other mentioned activities.

It is obvious that reading onboard increases most the probability of choosing a transport mode, while the lowest is for eating/drinking. Eating/drinking is not allowed onboard of public transport (Budapest public transport), and people are used to not eat or drink onboard to avoid penalties. Thus, their interest to eat/drink onboard is not that big interest. Writing shows negative impacts because to write onboard you need longer travel time and more comfortable environment. The study focused on trip time inside urban areas where the average trip time is around 20–30 min. The relative log odds of being in PT versus PAV will increase by 0.2049 if moving from the female to male. The relative log odds of being in PT versus PAV will increase by 0.36 if moving from the car ownership to people who do not own cars. Additionally, the males are more interested to use PT than female, and car ownerships are more interested to use PT than who do not own conventional cars. The probability of using PT for males is higher than the probability of using PAV. The probability of using PAV for car ownership is less than using PT.

Table 2 shows the predictive margins across multitasking onboard of PAV and PT. Six options of multitasking are examined across alternatives. In the case of PAV, the highest probability of choosing PAV is 58.25% when reading is available onboard, and the lowest probability of choosing PAV is 51.8% when writing is available onboard. The highest probability of choosing PT is 47.8% when reading is available. The lowest probability of choosing PT is 41.5% when writing onboard is available. It is shown in Table 2 that the probability of choosing PAV is higher than PT considering the availability of certain activities onboard (the margins of PAV are higher than the margins of PT). It is worth mentioning that the margins are significant at a confidence interval of 95%.

**Table 1.** Conditional logit-model estimates

| Alt. [a] | Attribute | Sub-attribute/level | Value | Std. error | z | P > \|z\| [a] |
|---|---|---|---|---|---|---|
| | Time | - | −0.0311 | 0.0064 | −4.78 | 0.000*** |
| | Cost | - | −0.0017 | 0.0004 | −3.99 | 0.000*** |
| | Multitasking options | Reading | 0.236 | 0.0927 | 2.55 | 0.011** |
| | | Writing | −0.0267 | 0.0986 | −0.27 | 0.786 |
| | | Talking | 0.0223 | 0.085 | 2.62 | 0.009*** |
| | | Using social media | 0.1903 | 0.0904 | 2.11 | 0.035** |
| | | Eating and drinking | 0.0049 | 0.094 | 0.05 | 0.95 |
| | | Others (doing nothing) | base | - | - | - |
| PT | Gender | Male | 0.2049 | 1532 | 1.34 | 0.181 |
| | | Female (base) | - | - | - | - |
| | Car ownership | No | 0.36 | 0.1596 | 2.25 | 0.024** |
| | | Yes (base) | - | - | - | - |
| | β0 | - | −0.4373 | 0.1227 | −3.56 | 0.000*** |
| PAV | Base | - | - | - | - | - |

Number of observations = 4366
Chi2(2) = 60.10
Log simulated-likelihood = −1467.5827
Prob > chi2 = 0.000
AIC = 2955.165
BIC = 3012.05
[a] Alt.: Alternative. *** $p < 0.01$, ** $p < 0.05$, * $p < 0.1$.

Table 3 shows the variation of margins with travel time changes. Three values of travel time are used here to see how margins are changed. In the case of PAV, the probability of selecting PAV is reduced when travel time is increased. For example, the margin of PAV is 58.8% at travel time of 20 min, and the margin is 51.3% at travel time of 30 min. What happens when the travel time of the opponent alternatives is changed? The probability of choosing PAV is increased when the travel time of the PT (i.e., opponent) is increased. For example, when the travel time of PT is increased from 20 min to 30 min, the probability of selecting PAV is increased from 51.5% to 58.9%. In the case of PT, the probability of selecting PT is reduced when travel time is increased. For example, the margin of PT is 48.5% at travel time of 20 min, and the margin is 41.1% at travel time of 30 min. What happens when the travel time of the opponent alternatives (i.e., PAV) is changed? The probability of choosing PT is increased when the travel time of the PAV is increased. For example, when the travel time of PAV is increased from 20 min to 30 min, the probability of selecting PT is increased from 41.2% to 48.7%. It is found that the change on the margins in the case of PAV is higher than PT, for example, changing the travel time from 20 min to 25 min the margin is decreased by 3.7% in the case of PT, while it is 3.8% in cased of PT, for the same travel time interval. Figure 4 shows the probability of choosing PAV and PT when travel time is changed. It is shown that travelers are more willing to use PAV over PT.

**Table 2.** Predictive margins across multitasking (Delta method)

| Alt | Attribute | Margin | Std. Error | z | P > \|z\| | 95% confidence interval | |
|---|---|---|---|---|---|---|---|
| PAV | Multitasking = reading | 0.582 | 0.023 | 25.72 | *** | 0.537 | 0.626 |
| | Multitasking = writing | 0.518 | 0.024 | 21.63 | *** | 0.471 | 0.565 |
| | Multitasking = talking | 0.578 | 0.022 | 25.79 | *** | 0.534 | 0.622 |
| | Multitasking = using social media | 0.571 | 0.022 | 26.05 | *** | 0.528 | 0.613 |
| | Multitasking = eating and drinking | 0.526 | 0.024 | 22.14 | *** | 0.479 | 0.572 |
| | Multitasking = others (doing nothing) | 0.525 | 0.024 | 22.28 | *** | 0.479 | 0.571 |
| PT | Multitasking = reading | 0.478 | 0.023 | 21.02 | *** | 0.434 | 0.523 |
| | Multitasking = writing | 0.415 | 0.024 | 17.27 | *** | 0.368 | 0.462 |
| | Multitasking = talking | 0.475 | 0.022 | 21.57 | *** | 0.432 | 0.518 |
| | Multitasking = using social media | 0.467 | 0.023 | 20.58 | *** | 0.423 | 0.512 |
| | Multitasking = eating and drinking | 0.423 | 0.023 | 18.62 | *** | 0.378 | 0.467 |
| | Multitasking = others (doing nothing) | 0.421 | 0.024 | 17.8 | *** | 0.375 | 0.468 |

Number of observations = 4366
*** $p < 0.01$, ** $p < 0.05$, * $p < 0.1$.

Table 4 shows the variation of margins with travel cost changes. Three values of travel cost are used to see how margins change. In the case of PAV, the probability of selecting PAV is reduced when travel cost is increased. For example, the margin of PAV is 58.2% at travel cost of 175 Ft, and the margin is 51.8% at travel cost of 300 Ft. What happens when the travel cost of the opponent alternatives is changed? The probability of choosing PAV is increased when the travel cost of the PT (i.e., opponent) is increased. For example, when the travel cost of PT is increased from 175 Ft to 300 Ft, the probability of selecting PAV is increased from 51.9% to 58.4%. In the case of PT, the probability of selecting PT is reduced when travel cost is increased. For example, the margin of PT is 48.1% at travel cost of 175 Ft, and the margin is 41.6% at travel cost 300 Ft. What happens when the travel cost of the opponent alternatives (i.e., PAV) is changed? The probability of choosing PT is increased when the travel cost of the PAV is increased. For

**Table 3.** Predictive margins across travel time (Delta method)

| Alt | At alternative (x) at travel time (i) | Margin | Std. error | z | P > \|z\| | 95% confidence interval | |
|---|---|---|---|---|---|---|---|
| PAV | PAV @ 20 min | 0.588 | 0.02 | 29.72 | *** | 0.549 | 0.626 |
| | PAV @ 25 min | 0.551 | 0.018 | 29.85 | *** | 0.514 | 0.587 |
| | PAV @ 30 min | 0.513 | 0.02 | 25.64 | *** | 0.474 | 0.552 |
| | PT @ 20 min | 0.515 | 0.02 | 25.84 | *** | 0.476 | 0.554 |
| | PT @ 25 min | 0.552 | 0.018 | 29.89 | *** | 0.516 | 0.589 |
| | PT @ 30 min | 0.589 | 0.02 | 29.57 | *** | 0.55 | 0.628 |
| PT | PAV @ 20 min | 0.412 | 0.02 | 20.86 | *** | 0.374 | 0.451 |
| | PAV @ 25 min | 0.449 | 0.018 | 24.36 | *** | 0.413 | 0.486 |
| | PAV @ 30 min | 0.487 | 0.02 | 24.34 | *** | 0.448 | 0.526 |
| | PT @ 20 min | 0.485 | 0.02 | 24.37 | *** | 0.446 | 0.524 |
| | PT @ 25 min | 0.448 | 0.018 | 24.23 | *** | 0.411 | 0.484 |
| | PT @ 30 min | 0.411 | 0.02 | 20.6 | *** | 0.372 | 0.45 |

Number of observations = 4366
*** $p < 0.01$, ** $p < 0.05$, * $p < 0.1$.

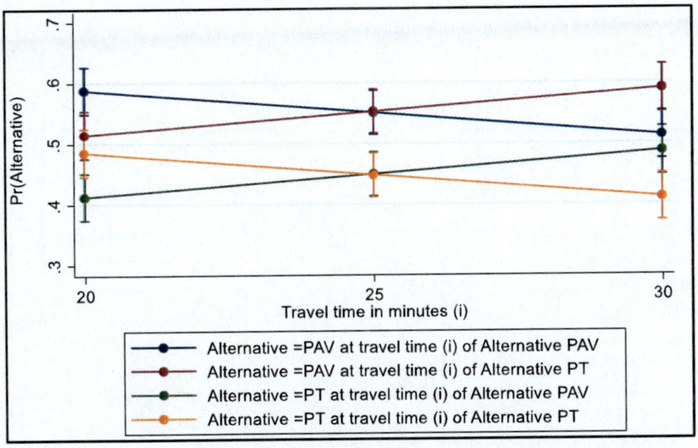


**Fig. 4.** Prediction margin based on the changes in travel time

example, when the travel cost of PAV is increased from 175 Ft to 300 Ft, the probability of selecting PT is increased from 41.8% to 48.2%. It is found that the change on the margins in the case of PT is higher than PT, for example, changing the travel cost from 175 Ft minutes to 250 Ft, the margin is decreased by 3.2% in the case of PAV, while it is 3.3% in case of PT, for the same travel time interval. Figure 5 shows the probability of choosing PAV and PT when travel cost is changed. It is shown that travelers are more willing to use PAV over PT.

**Table 4.** Predictive margins across travel cost (Delta method)

| Alt | At alternative (x) at travel time (i) | Margin | Std. error | z | P > \|z\| | 95% confidence interval | |
|---|---|---|---|---|---|---|---|
| PAV | PAV @ 175 Ft. | 0.582 | 0.02 | 29.04 | *** | 0.543 | 0.622 |
| | PAV @ 250 Ft. | 0.55 | 0.018 | 29.87 | *** | 0.514 | 0.586 |
| | PAV @ 300 Ft. | 0.518 | 0.02 | 25.95 | *** | 0.479 | 0.557 |
| | PT @ 175 Ft. | 0.519 | 0.02 | 26.21 | *** | 0.481 | 0.558 |
| | PT @ 250 Ft. | 0.552 | 0.018 | 29.94 | *** | 0.516 | 0.588 |
| | PT @ 300 Ft. | 0.584 | 0.02 | 28.91 | *** | 0.544 | 0.623 |
| PT | PAV @ 175 Ft. | 0.418 | 0.02 | 20.83 | *** | 0.378 | 0.457 |
| | PAV @ 250 Ft. | 0.45 | 0.018 | 24.4 | *** | 0.414 | 0.486 |
| | PAV @ 300 Ft. | 0.482 | 0.02 | 24.15 | *** | 0.443 | 0.521 |
| | PT @ 175 Ft. | 0.481 | 0.02 | 24.24 | *** | 0.442 | 0.519 |
| | PT @ 250 Ft. | 0.448 | 0.018 | 24.31 | *** | 0.412 | 0.484 |
| | PT @ 300 Ft. | 0.416 | 0.02 | 20.61 | *** | 0.377 | 0.456 |

Number of observations = 4366
*** $p < 0.01$, ** $p < 0.05$, * $p < 0.1$.

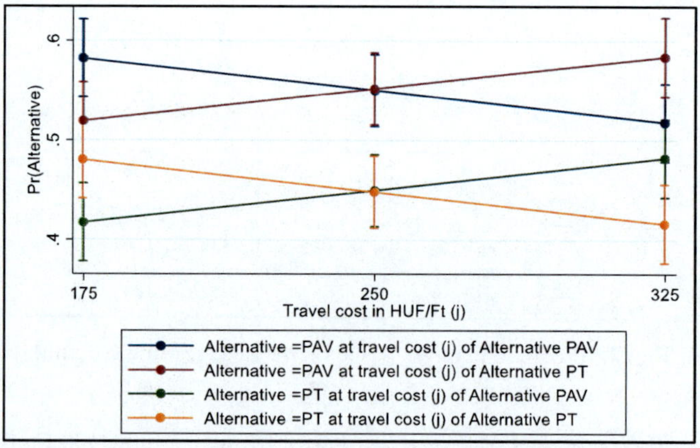


**Fig. 5.** Prediction margin based on the changes in travel cost

Therefore, the characteristics of alternatives influence travel behavior. The type of multitasking, travel time, and travel cost are important factors that influence travel behavior, as demonstrated in Fig. 4 and Fig. 5. The study demonstrates that travelers are more willing to use PAV over PT. Furthermore, the research demonstrates that all multitasking options, with the exception of writing, have a positive impact on the utility of travel. Increasing travel time decreases the probability of choosing a transport mode, which decreasing the travel cost increases the probability of choosing a transport mode.

Limitations of the study are summarized in the ability of respondents to take a decision about a transport mode that is not existed in the market yet. Text and photos of the new technology (i.e., PAV) are given to respondents. Additionally, the sample size is normally insufficient to include other variables in the model such as income, trip purpose, and other variables. The study is conducted in Budapest where people record their main trips inside the urban areas, where the trip is relatively short. Thus, the study is applicable in urban areas where traveling is relatively short.

Few studies are conducted in Hungary that focus on the acceptability of people to the new technology (PAV and SAV) [14, 15]. Those studies do not include the multitasking onboard as attributes. The study the differences in the acceptability across several countries as well as the changes in acceptability across sociodemographic groups.

The recommendations are expanding the study to cover different countries to show the variations across countries. Besides, including different trips length (i.e., intracity and intercity) to estimate the utility of each onboard activity per trip length (where the divisions of multitasking might be varied according to new research objectives) is also recommended future work. Including SAV in further study to estimate the impact of onboard activities on travel behavior when PAV and SAV are introduced, is a recommended study in the future.

## 5 Conclusion

The travel behavior of people is changing with the advancement of technology. AV is a promising technology that will change the behavior of traveling onboard because the machine takes the role of driving instead of the human driver. In this study, the behavior of travelers onboard of PAV and PT is analyzed by applying a discrete choice modeling approach. The main daily trip inside urban areas is studied, where trip time is relatively short. An SP is conducted in Budapest, Hungary, where DCE is designed and distributed randomly. Three variables are studied, travel time, travel cost, and multitasking onboard. The study uses a list of multitasking, such as reading, writing, talking, using social media, eating-drinking, and doing other things. The conditional Logit model is applied, where the characteristics of PAV and PT are considered. As a result, a transport choice model is developed which combines trip time, trip cost, multitasking availability, car ownership, and gender variables. The results demonstrate that the trip time and the trip cost show negative impact on the utility of traveling which is consistent with random utility theory. While the onboard activities show varied results on the utility of travel. For example, writing onboard adds a negative utility to the travel utility which impacts choosing a transport mode, while other activities add a positive utility. Moreover, different results are presented across gender and car ownership. This demonstrates the impact of car ownership and gender groups on the selection of a transport mode, where the probability of using PAV and PT is affected by the gender and car ownership status of traveler. In conclusion, the results show that people are more likely to choose PAV over PT inside urban areas, where the margins show a larger modal share of PAV than PT at the same trip time and trip cost.

**Acknowledgment.** The research reported in this paper and carried out at the Budapest University of Technology and Economics has been supported by the National Research Development and Innovation Fund (TKP2020 Institution Excellence Subprogram, Grant No. BME-IE-MISC) based on the charter of bolster issued by the National Research Development and Innovation Office under the auspices of the Ministry for Innovation and Technology.

## References

1. Cascetta, E.: Random utility theory. In: Transportation Systems Analysis, pp. 89–167. Springer (2009)
2. Belenky, P.: The Value of Travel Time Savings: Departmental Guidance for Conducting Economic Evaluations, Revision 2. Department of Transportation (2011)
3. Mackie, P.J., Jara-Dıaz, S., Fowkes, A.: The value of travel time savings in evaluation. Transp. Res. Part E Logist. Transp. Rev. **37**(2–3), 91–106 (2001). https://doi.org/10.1016/S1366-5545(00)00013-2
4. Varghese, V., Jana, A.: Impact of ICT on multitasking during travel and the value of travel time savings: empirical evidences from Mumbai India. Travel Behav. **12**, 11–22 (2018). https://doi.org/10.1016/j.jtrangeo.2012.02.007
5. Steck, F., et al.: How autonomous driving may affect the value of travel time savings for commuting. Transp. Res. Record J. Transp. Res. Board **2672**(46), 10 (2018). https://doi.org/10.1177/0361198118757980
6. Kouwenhoven, M., de Jong, G.: Value of travel time as a function of comfort. J. Choice Model. **28**, 97–107 (2018). https://doi.org/10.1016/j.jocm.2018.04.002
7. Cherlow, J.R.: Measuring values of travel time savings. J. Consum. Res. **7**(4), 360–371 (1981). https://doi.org/10.1086/208826
8. Litman, T.: Valuing transit service quality improvements. J. Public Transp. **11**(2), 3 (2008). https://doi.org/10.5038/2375-0901.11.2.3
9. Coppola, P., Esztergár-Kiss, D.: Autonomous Vehicles and Future Mobility Elsevier (2019)
10. Ettema, D., Verschuren, L.: Multitasking and value of travel time savings. Transp. Res. Record J. Transp. Res. Board **2010**(1), 19–25 (2007). https://doi.org/10.3141/2010-03
11. Malokin, A., Circella, G., Mokhtarian, P.L.: How do activities conducted while commuting influence mode choice? Using revealed preference models to inform public transportation advantage and autonomous vehicle scenarios. Transp. Res. Part A Policy and Pract. **124**, 82–114 (2019). https://doi.org/10.1016/j.tra.2018.12.015
12. Rhee, K.-A., et al.: Analysis of effects of activities while traveling on travelers' sentiment. Transp. Res. Record J. Transp. Res Board **2383**(1), 27–34 (2013). https://doi.org/10.3141/2383-04
13. Mokhtarian, P.L., Papon, F., Goulard, M., Diana, M.: What makes travel pleasant and/or tiring? An investigation based on the French National Travel Survey. Transportation **42**(6), 1103–1128 (2014). https://doi.org/10.1007/s11116-014-9557-y
14. Etzioni, S., et al.: Modeling cross-national differences in automated vehicle acceptance. Sustainability **12**(22), 9765 (2020). https://doi.org/10.3390/su12229765
15. Polydoropoulou, A., et al.: Who is willing to share their AV? Insights about gender differences among seven countries. Sustainability **13**(9), 4769 (2021). https://doi.org/10.3390/su13094769

16. Berliner, R.M., et al., Travel-based multitasking: modeling the propensity to conduct activities while commuting. In: Transportation Research Board 94th Annual Meeting Washington, DC (2015)
17. Banerjee, I., Kanafani, A.: The value of wireless internet connection on trains: implications for mode-choice models. In: UC Berkeley, U.o.C.T. Center, Berkeley (2008)
18. Pudāne, B., et al.: A time-use model for the automated vehicle-era. Transp. Res. Part C Emerg. Technol. **93**, 102–114 (2018). https://doi.org/10.1016/j.trc.2018.05.022
19. Keseru, I., Macharis, C.: Travel-based multitasking: review of the empirical evidence. Transp. Rev. **38**(2), 162–183 (2018). https://doi.org/10.1080/01441647.2017.1317048
20. Pawlak, J.: Travel-based multitasking: review of the role of digital activities and connectivity. Transp. Rev. **40**(4), 429–456 (2020). https://doi.org/10.1080/01441647.2020.1728418
21. Khaloei, M., Ranjbari, A., Mackenzie, D.: Analyzing the Shift in Travel Modes' Market Shares with the Deployment of Autonomous Vehicle Technology (2019)
22. Singleton, P.A.: Discussing the "positive utilities" of autonomous vehicles: will travellers really use their time productively? Transp. Rev. **39**(1), 50–65 (2019). https://doi.org/10.1080/01441647.2018.1470584
23. Zhong, H., et al.: Will autonomous vehicles change auto commuters' value of travel time? Transp. Res. Part D Transp. Environ. **83,** 102303 (2020). https://doi.org/10.1016/j.trd.2020.102303
24. Becker, G.S.: A theory of the allocation of time. Econ. J. **75**, 493–517 (1965). https://doi.org/10.2307/2228949
25. Jara-Díaz, S.R.: Allocation and valuation of travel time savings. Handbooks Transp. **1**, 303–319 (2000). https://doi.org/10.1108/9780857245670-018
26. Horni, A., Nagel, K., Axhausen, K.W.: The Multi-Agent Transport Simulation MATSim, p. 620, Ubiquity Press, London (2016)
27. Hamadneh, J., Esztergár-Kiss, D.: Impacts of shared autonomous vehicles on the travelers' mobility. In: 2019 6th International Conference on Models and Technologies for Intelligent Transportation Systems (MT-ITS), Poland. IEEE (2019). https://doi.org/10.1109/MTITS.2019.8883392
28. Cirillo, C., Axhausen, K.W.: Evidence on the distribution of values of travel time savings from a six-week diary. Transp. Res. Part A Policy Pract. **40**(5), 444–457 (2006). https://doi.org/10.1016/j.tra.2005.06.007
29. Xumei, C., Qiaoxian, L., Guang, D.: Estimation of travel time values for urban public transport passengers based on SP survey. J. Transp. Syst. Eng. Inf. Technol. **11**(4), 77–84 (2011). https://doi.org/10.1016/S1570-6672(10)60132-8
30. Athira, I., et al.: Estimation of value of travel time for work trips. Transp. Res. Procedia **17**, 116–123 (2016). https://doi.org/10.1016/j.trpro.2016.11.067
31. Janssen, C.P., Kenemans, J.L.: Multitasking in autonomous vehicles: ready to go? In: 3rd Workshop on User Experience of Autonomous Vehicles at AutoUI 2015. ACM Press, Nottingham (2015)
32. Ortega, J., et al.: Simulation of the daily activity plans of travelers using the park-and-ride system and autonomous vehicles: work and shopping trip purposes. Appl. Sci. **10**(8), 2912 (2020). https://doi.org/10.3390/app10082912
33. Ian Wallis Associates Ltd.: Car passenger valuations of quantity and quality of time savings. In NZ Transport Agency Research Report 551, NZ Transport Agency: New Zealand, p. 64 (2014)
34. Simoni, M.D., et al.: Congestion pricing in a world of self-driving vehicles: an analysis of different strategies in alternative future scenarios. Transp. Res. Part C Emerg. Technol. **98**, 167–185 (2018). https://doi.org/10.1016/j.trc.2018.11.002

35. Kolarova, V., Steck, F., Bahamonde-Birke, F.J.: Assessing the effect of autonomous driving on value of travel time savings: a comparison between current and future preferences. Transp. Res. Part A Policy Pract. **129**, 155–169 (2019). https://doi.org/10.1016/j.tra.2019.08.011
36. Bozorg, S.L., Ali, S.M.: Potential Implication of Automated Vehicle Technologies on Travel Behavior and System Modeling (2016)
37. Gurumurthy, K.M., Kockelman, K.M.: Modeling Americans' autonomous vehicle preferences: A focus on dynamic ride-sharing, privacy & long-distance mode choices. Technol. Forecast. Soc. Change **150,** 119792 (2020). https://doi.org/10.1016/j.techfore.2019.119792
38. Lavieri, P.S., Bhat, C.R.: Modeling individuals' willingness to share trips with strangers in an autonomous vehicle future. Transp. Res. Part A Policy Pract. **124**, 242–261 (2019). https://doi.org/10.1016/j.tra.2019.03.009
39. Wadud, Z., Huda, F.Y.: The potential use and usefulness of travel time in fully automated vehicles. In: Transportation Research Board 97th Annual Meeting. Washington, DC (2018)
40. Hauber, A.B., et al.: Statistical methods for the analysis of discrete choice experiments: a report of the ISPOR Conjoint Analysis Good Research Practices Task Force. Value Health **19**(4), 300–315 (2016). https://doi.org/10.1016/j.jval.2016.04.004
41. Schmitz, C.: LimeSurvey: an open source survey tool. LimeSurvey Project Hamburg, Germany (2012). http://www.limesurvey.org
42. Lancsar, E., Louviere, J.: Conducting discrete choice experiments to inform healthcare decision making. Pharmacoeconomics **26**(8), 661–677 (2008). https://doi.org/10.2165/00019053-200826080-00004
43. ChoiceMetrics: Ngene 1.0 User manual and reference guide: the cutting edge in experimental design (2014)
44. Hamadneh, J., Esztergár-Kiss, D.: Potential travel time reduction with autonomous vehicles for different types of travellers. Promet-Traffic Transp. **33**(1), 61–76 (2021). https://doi.org/10.7307/ptt.v33i1.3585
45. Hoffman, S.D., Duncan, G.J.: Multinomial and conditional logit discrete-choice models in demography. Demography **25**(3), 415–427 (1988). https://doi.org/10.2307/2061541
46. Stata Corporation: Stata Base Reference Manual: Release, vol.16. Stata Corporation, College Station (2019)

# Evolution of Probabilistic Characteristics in the Train Traffic Process

Boris Davydov[1], Vladimir Chebotarev[2], and Kseniya Kablukova[2](✉)

[1] Far Eastern State Transport University, Khabarovsk, Russia
[2] Computing Center of the Far Eastern Branch of the Russian Academy of Sciences, Khabarovsk, Russia
{vladimir.ch,kseniya_kablukova}@ccfebras.ru

**Abstract.** Trajectory elements of train movement, such as departure and running times, are subject to random influences, which can lead to disruption of the arrivals. Analysis of the corresponding probability distributions and their changes, which arise depending on intermediate stations, makes it possible to identify the degree of influence of various factors on the movement process, as well as to assess the quality of train traffic control. This article provides an approach for calculating the distribution of the arrival time deviations and investigates the evolution of this distribution when trains travel on an extended section of a main railway line. Historical data on the scheduled freight train traffic at the Russian Railways are used to verify the proposed model.

**Keywords:** Train traffic · Stochastic model · Deviations from the schedule · Statistical analysis · Operational times distribution

## 1 Introduction

Train traffic is affected by a number of random factors. Such factors can be unfavorable weather conditions, time of day, technical failures of facilities, etc. The result of their impact are disturbances of the train schedule. It is necessary to operational adjusting the schedule to ensure the continued normal functioning of the railway system.

The original schedule includes the departure and arrival times at fixed points of the route, that is, at the stations. Random factors can lead to the fact that the departure/arrival times at the station become random variables. In a simplified way, we can say that in the present work the following problem is solved: on the assumption that the distributions of departure deviation from a certain station, travel time and dwell time are known, find the distributions of arrival deviation at subsequent stations. The solution of the problem will allow both predicting arrival deviations and adjusting the parameters of the train traffic process.

Let us imagine a railway line with both the origin and the terminal stations. It is assumed that there are a number of intermediate stations between them. The schedule of arrivals and departures for all stations is known (for the terminal station, only the planned arrival time). Let the train departs from the sourcing station. As a result of


G. Sierpiński (Ed.): TSTP 2021, LNNS 352, pp. 101–109, 2022.
https://doi.org/10.1007/978-3-030-91156-0_8

random factors' impact, the schedule may be disturbed. The main goal of the research is to derive an algorithm for finding the probability distribution of the arrival time deviation at the terminal station.

This article deals with the case of only one intermediate station. The random arrival time deviation $\xi_3$ from the timetable at the terminal station depends on three random variables: $\delta_1$ is the departure time deviation at the sourcing station, $\tau_2$ is the dwell time at the intermediate station, and $\rho$ is the total travel time between the sourcing and terminal stations. So the desired distribution is determined by the joint distribution of three input random variables: $\delta_1$, $\tau_2$, and $\rho$. Generally speaking, the search for the joint distribution of three random variables is a difficult problem, given that they can be stochastically dependent. In proposed model, we impose the condition that $\delta_1$, $\tau_2$, and $\rho$ are mutually independent. On the one hand, this helps us to solve the assigned task in the particular case considered, and on the other hand, the indicated model assumption is quite often fulfilled in practice, which is confirmed by the results of the real statistical analysis given in Sect. 6.

The article is structured as follows. Section 2 reviews the literature on the problem of modeling train traffic. Section 3 is devoted to the description of the stochastic model of train movement and the basic assumptions used to construct the arrival deviation distribution. Section 4 discusses the issue of approximating the input distributions used in modeling. In Sect. 5 there is derived a formula for the distribution density of the deviation upon arrival at the terminal station. Section 6 verifies the stochastic model and checks the adequacy of the initial assumptions. Section 7 contains conclusions.

## 2 Literature Review

Models for predicting the train travel parameters can be divided, depending on their ability to cope with emerging uncertainties, into deterministic and stochastic. Deterministic models assume complete knowledge of the future development of the situation. Stochastic models assign a probability distribution to each event in order to simulate the uncertainty of its implementation. Also, the presented approaches are divided into static and dynamic, depending on how they use information in real time to update their forecasts. While static prediction models are based on autonomously computed estimates and parameters, dynamic models are updated in real time.

In deterministic forecasting, train traffic is modeled as a discrete dynamic system of events, represented by a directed acyclic graph. System events, as a rule, include only station events (arrivals, departures). The operation time in most approaches is calculated using theoretical estimates. Travel times are calculated using a train traffic equation that takes into account the dynamic properties of rolling stock and network infrastructure. They are assumed to be fixed and do not change depending on the information received in real time from the monitoring system. Dwell times are predicted with fixed estimates. An example of such a deterministic static forecasting model was implemented in [1]. After each update of the train position, the critical path algorithm predicts the time of all events on the graph.

Another example is presented in [2]. It describes a prediction system for a high-speed network in Japan. Based on information from the monitoring system, the train trajectory to the next station is simulated in real time mode. The delay is then extrapolated for the next stations. D'Ariano et al. [3] presented a graph-based model for real-time rescheduling. Accurate prediction of future train coordinates was an important requirement for assessing the effect of proposed timetable adjustments. In an improved approach, estimates of the travel times of trains involved in route conflicts are updated online to simulate braking and re-acceleration [4].

The availability of historical data on the executed train traffic has stimulated the creation of deterministic forecasting models in the other direction. Statistical analysis and data mining techniques can be used to assess the dependence of operating times on factors such as delay and peak hours [5]. The operating time is estimated online depending on the information received by the monitoring system. This method was used in [6]. The authors presented a method for predicting the operating time of the event by calculating the critical path from the last realized event. Kecman and Goverde [7] extended this approach by simulating train traffic at the block section level.

Deterministic approaches focus on explaining the uncertainty of the operation time. Although more advanced data-driven models can explain a large percentage of its variability using the values of the independent variables, some degree of variability is still unresolved. Another important aspect of using the presented models for real-time forecasting is that their accuracy largely depends on the spatial and temporal updates received from the monitoring systems. Consequently, to obtain accurate predictions, a detailed model of railway traffic with frequent updates of the trains' coordinates is required.

Stochastic models were mainly developed to analyze the stability and robustness of schedules [8]. Buker and Seybold [9] modeled delays as random variables described by suitable distribution functions and applied analytical methods to calculate the delay propagation process. An approach to delay propagation based on processed data from train event recorders was presented in [10]. A method based on asynchronous simulation of the movement of individual trains using probabilistic distributions of the traffic parameters was proposed in [11]. Meester and Muns [12] formalized the model presented in [13] as a stochastic event graph. The distributions of random variables in the absence of conflicts are considered known, and the dependencies are calculated as a mixture of the corresponding distributions. A number of performance indicators are derived as a linear combination of delay distributions. Yuan and Hansen [14] described a detailed analytical stochastic delay propagation model for complex station sections. Analytical formulas are given for calculating the secondary delays of departing and arriving trains. The conditional distributions of arrival and departure times are calculated.

The described stochastic approaches are inherently static. They do not take into account the effect of information received in real time from the monitoring system on reducing the uncertainty of future events. The first stochastic approach, which to some extent uses the current conditions to reduce uncertainty, is presented in [15]. The authors have created a stochastic graph model to predict delays. This approach is suitable for online applications that frequently update train location data. The travel times are modeled as random variables. Several theoretical distributions have been proposed for travel times. Despite the fact that up-to-date information on train departure times can be used to adjust estimates of subsequent travel times, the proposed approach does not provide an explicit conditional probability between travel time and departure delay.

## 3 The Model for Arrival Time Deviation Forming

Consider the single train traffic along the railway section. The train departs from some station $S_1$, overcomes a number of tracks and stations and arrives at station $S_n$, for which it is necessary to determine the arrival deviation distribution. Let us describe the values involved in this process. In the description, we will restrict ourselves to three stations $S_1$, $S_2$ and $S_3$.

The train departs from $S_1$ at the time $d_1+\delta_1$, where $d_1$ denotes the scheduled departure time, $\delta_1$ denotes the random departure deviation from schedule. Further, the movement along the track between $S_1$ and $S_2$ occurs, which is characterized by the random running time $\rho_1$. At station $S_2$ there is a stop with dwell time $\tau_2$. The movement resumes after stop. The travel time between $S_2$ and $S_3$ is denoted by $\rho_2$. The actual arrival time at the station $S_3$ is equal to $a_3 + \xi_3$, where $a_3$ is the planned arrival time, and $\xi_3$ is the random arrival deviation. The arrival deviation $\xi_3$ is the investigated random variable. Figure 1 depicts the diagram of the arrival time $a_3 + \xi_3$ formation. The described approach can be generalized to an arbitrary number of stations and tracks.

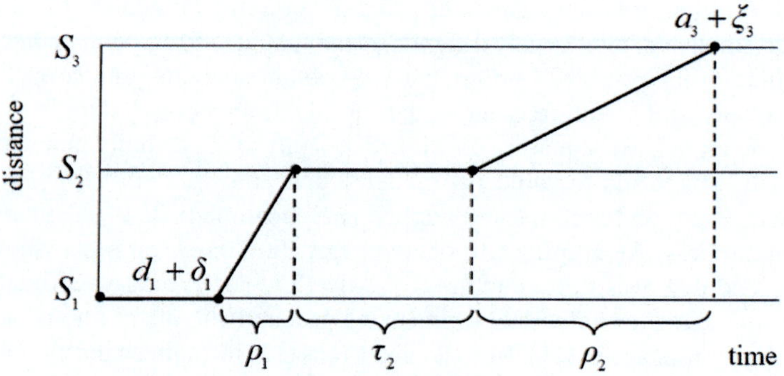


**Fig. 1.** The diagram of the arrival time $a_3 + \xi_3$ formation

According to the diagram in Fig. 1, the arrival time formation is a process of summing up the random variables characterizing the elements of the trajectory in the "time – distance" coordinates horizon. These elements are the departure time from the beginning station, travel times between stations, as well as the dwell times at stations. The number of intermediate stations can be arbitrary.

According to Fig. 1, the arrival time at $S_3$ can be written as:

$$a_3 + \xi_3 = d_1 + \delta_1 + \rho_1 + \tau_2 + \rho_2, \quad (1)$$

whence

$$\xi_3 = d_1 - a_3 + \delta_1 + \rho_1 + \tau_2 + \rho_2. \quad (2)$$

In turn, it follows from (2) that

$$\xi_3 = \delta_1 + \tau_2 + \rho + d_1 - a_3, \quad (3)$$

where $\rho = \rho_1 + \rho_2$. Equality (3) plays a key role in the derivation of the formula for the probability density function of $\xi_3$ in Sect. 5.

In what follows, the distribution functions of random variables $\xi_3$, $\delta_1$, $\rho_j$ $(j = 1, 2)$, $\rho$, $\tau_2$ will be denoted by $U_3(t)$, $F_1(t)$, $L_j(t)$, $L(t)$ and $G_2(t)$, respectively, i.e.

$$U_3(t) = \mathrm{P}(\xi_3 < t), \;\; F_1(t) = \mathrm{P}(\delta_1 < t), \;\; L_j(t) = \mathrm{P}\big(\rho_j < t\big),$$
$$L(t) = \mathrm{P}(\rho < t), \;\; G_2(t) = \mathrm{P}(\tau_2 < t).$$

We assume that the densities of these random variables exist. Let us denote them by the corresponding lowercase letters: $u_3(t), f_1(t), l_j(t), l(t)$ and $g_2(t)$.

Let us proceed to the description of the statistical data used and approximations of the input random variables' distributions.

## 4 Distributions of Train Operating Times

The main purpose of this article is to construct the distribution of arrival deviations at a fixed station, taking as the initial data the distribution of random variables characterizing the previous movement.

As the initial data, statistical data on the scheduled freight trains obtained at one of the sections of the Trans–Siberian Railway, are taken. This Railway is one of the main railways of Russia. It is distinguished by the presence of intensive mixed freight and passenger traffic.

The train traffic on a railway section of 655 km is considered. This section is adjacent to the main terminal (Khabarovsk). A preliminary analysis [16] showed that the statistical characteristics of the freight trains' traffic along this line change depending on the time of day. In addition, it is found out that there is no correlation between operating times in the morning hours. A similar assumption has been put forward and will be verified when analyzing data on the traffic of the morning trains.

By using the statistical data obtained on the considered railway section, let us determine the types of distributions of the input random variables: departure deviation $\delta_1$, dwell time $\tau_2$ and travel time $\rho$ along the section.

In previous research it was shown (see [17]) that deviations from the schedule and operating times are in many cases well described by an exponential distribution with a shift, which is equal to $b$:

$$f(x) = I(x > b)\lambda \exp\{-\lambda(x - b)\}, \tag{4}$$

where $\lambda > 0, b \in \mathbb{R}$ are the distribution parameters. In (4) and below, we use the notation

$$I(x \in A) = \begin{cases} 1 \; if \;\; x \in A, \\ 0 \; if \;\; x \in \mathbb{R} \backslash A, \end{cases}$$

where $A$ is an arbitrary set on the real line $\mathbb{R}$.

Table 1 contains data of the input random variables' distributions for the morning trains. The type and parameters of each probability density function are matched with the corresponding real-world data histogram.

The unit of measurement of the considered random variables is hours.

**Table 1.** The distributions of input operating times

| Random variable | Histogram and its approximating probability density function | Type and parameters of the approximating distribution |
|---|---|---|
| $\delta_1$ departure time deviation | 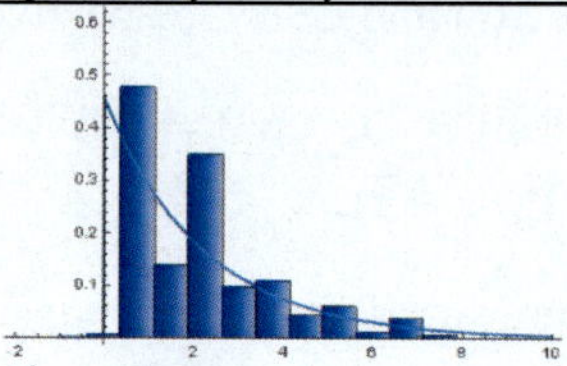 | Distribution with density (4), where $\lambda = 0.4585$, $b = 0$ |
| $\tau_2$ dwell time | 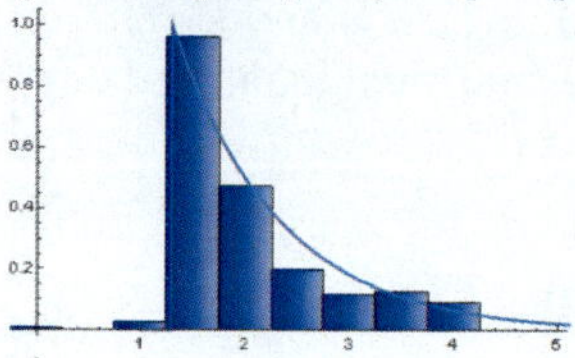 | Distribution with density (4), where $\lambda = 1$, $b = 1.3$ |
| $\rho$ travel time | 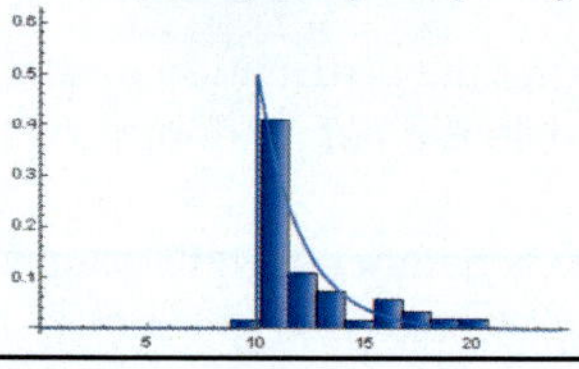 | Distribution with density (4), where $\lambda = 0.5$, $b = 10$ |

## 5 Derivation of the Arrival Deviation Distribution

Let us search for the probability distribution of the output random variable $\xi_3$, namely, the arrival deviation from the schedule at the terminal station of the examined section. The input random variables are $\delta_1$, $\tau_2$ and $\rho$, as indicated in Sect. 4. The joint distribution of these three random variables is unknown. However, based on the analysis of similar statistics on freight trains, an assumption about their mutual independence was suggested. By applying this assumption, let us check how well the resulting probability density function $u_3(t)$ will agree with the real data on the distribution of the random variable $\xi_3$, collected on the Russian Railways.

Random variables $\delta_1$, $\tau_2$ and $\rho$ are included in formula (3) as summands of a sum. It is known (see [18, P. 272]) that the probability density function of the sum of independent random variables is a convolution of the densities of these ones. The convolution (denoted by the symbol $*$) of two densities $\varphi_1(t)$ and $\varphi_2(t)$ is understood as the following integral

$$(\varphi_1 * \varphi_2)(t) = \int_{-\infty}^{\infty} \varphi_1(t-x)\varphi_2(x)dx. \tag{5}$$

Thus, using the representation (3), we arrive at the main result: the density function $u_3(t)$ of the random variable $\xi_3$ is equal to the convolution of the three input densities at the point $(t + a_3 - d_1)$:

$$\begin{aligned} u_3(t) &= (f_1 * g_2 * l)(t + a_3 - d_1) = ((f_1 * g_2) * l)(t + a_3 - d_1) \\ &= \int_{-\infty}^{\infty} (f_1 * g_2)(t + a_3 - d_1 - x)l(x)dx, \end{aligned} \tag{6}$$

where $(f_1 * g_2)(t + a_3 - d_1 - x) = \int\limits_{-\infty}^{\infty} f_1(t + a_3 - d_1 - x - y)g_2(y)dy$.

The type and parameters of the input densities are described in Table 1. According to (6), it is necessary to determine what the convolution of three exponential distributions with shifts is equal to.

Let $v_i(x) = I(x > b_i)\lambda_i \exp\{-\lambda_i(x - b_i)\}$, where $i = 1, 2, 3$, $\lambda_i > 0$, $b_i \in \mathbb{R}$ are the distribution parameters that do not coincide with each other. The convolution of the two distributions $v_1(x)$ and $v_2(x)$ is determined as follows:

$$(v_1 * v_2)(x) = I(x > b_1 + b_2)\frac{\lambda_1\lambda_2}{\lambda_1 - \lambda_2}\big[\exp\{-\lambda_2(x - b_1 - b_2)\} \\ - \exp\{-\lambda_1(x - b_1 - b_2)\}\big].$$

The convolution of the three exponential distributions with shifts has the form of the following function:

$$(v_1 * v_2 * v_3)(x) = I(x > b_1 + b_2 + b_3)\frac{\lambda_1\lambda_2\lambda_3}{\lambda_1 - \lambda_2} \\ \times\left[\frac{\exp\{-\lambda_2(x - b_1 - b_2 - b_3)\}}{\lambda_3 - \lambda_2} - \frac{\exp\{-\lambda_3(x - b_1 - b_2 - b_3)\}}{\lambda_3 - \lambda_2}\right. \\ \left. - \frac{\exp\{-\lambda_1(x - b_1 - b_2 - b_3)\}}{\lambda_3 - \lambda_1} + \frac{\exp\{-\lambda_3(x - b_1 - b_2 - b_3)\}}{\lambda_3 - \lambda_1}\right]. \tag{7}$$

It should be noted that the right-hand side of equality (7) is non-negative, since the convolution of non-negative functions is non-negative.

## 6 Verification of the Proposed Model

In the previous section, it was shown that the density function of the random variable $\xi_3$ is determined by the formula (6). Given the received estimates of parameters

$$\lambda_1 = 0.4585,\ b_1 = 0,\ \lambda_2 = 1,\ b_2 = 1.3,\ \lambda_3 = 0.5,\ b_3 = 10, \tag{8}$$

Let us check how well density function (6) with the indicated parameters agrees with the real data on the arrival deviations of container trains. Calculations are carried out according to the formula (7) at $x = t + a_3 - d_1$. The value $a_3 - d_1$ is a constant and equals 12.2 h.

Figure 2 depicts a graph of the probability density function $u_3(t)$ plotted in one coordinate system with the histogram for arrival deviations.

Given a sample $\{x_j\}_{j=1}^n$ from the distribution of a random variable $\xi_3$ of volume $n = 221$, wherein $\min x_j = 0.5$, $\max x_j = 11.5$. The number of intervals in the interval variation series is taken as $k = 11$.

The statistical hypothesis that this sample was taken from the distribution with density (6) with considering (7), in which the parameters are determined according to (8), should be checked.

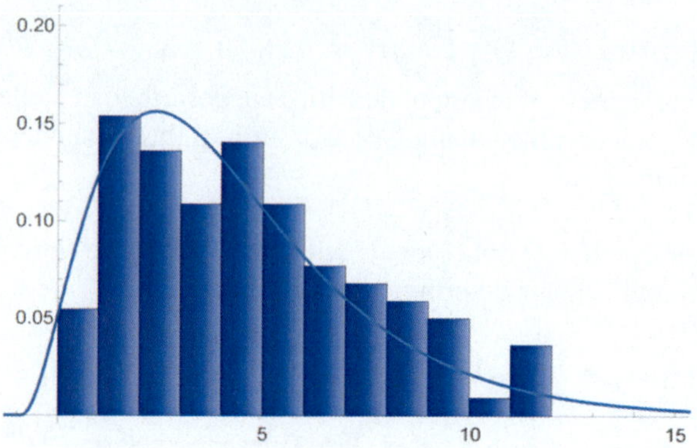


**Fig. 2.** Graph of the function $u_3(t)$ and the corresponding histogram

To test the hypothesis, we will use the chi-squared test, which is based on the following statistics

$$\chi^{2*} = \sum_{j=1}^{k} \frac{\left(n_j - np_j\right)^2}{np_j}, \tag{9}$$

where $n_j$ denotes the number of sample values that fall into the interval with index $j$, $p_j$ denotes the hypothetical probability of the random variable $\xi_3$ falling into this interval. Since the parameters of the hypothetical distribution are fully determined, according to the Pearson-Fisher theorem, the statistics $\chi^{2*}$ (provided that the hypothesis is correct) has a distribution close to the chi-squared distribution with $r = k - 1 = 10$ degrees of freedom.

Since the calculated value of statistics (9) is $\chi^{2*} = 15.41$, and it is less than the critical value $\chi^2_{0.05} = 18.31$, then the hypothesis formulated does not contradict the experimental data according to Pearson's chi-squared test with a significance level 0.05.

Based on this, we can conclude that the assumption about mutual independence of the input random variables (departure deviation, travel time and dwell time) is justified, which led to formula (6).

## 7 Conclusions

In this paper, we have developed a model described the evolution of the arrival deviation distribution at the terminal station. The model is based on the mutual independence assumption of the input random variables, which include the departure deviation, travel time and dwell time. This assumption makes it possible to derive a formula for the arrival deviation distribution. The derived formula (6) was tested on an array of statistical data collected on the scheduled freight traffic. Since this density formula does not contradict with the experimental data according to Pearson's chi-squared test with a significance level 0.05, then the model assumption of mutual independence is justified.

Our further research will be devoted to a more difficult case, when the assumption about the independence of the input random variables: $\delta_1$, $\rho$ and $\tau_2$ is not satisfied, and formula (6) contradicts statistical data.

## References


1. Dolder, U., Krista, M., Voelcker, M.: RCS – Rail Control System – Realtime train run simulation and conflict detection on a net wide scale based on updated train positions. In: 3rd International Seminar on Railway Operations Modelling and Analysis (RailZurich 2009), Zurich, pp. 1–15 (2009)
2. Fukami, K., Yamamoto, H.: A new diagram forecasting system for the TokaidoSanyo Shinkansen. In: World Congress on Railway Research, Koln, pp. 1–6 (2001)
3. D'Ariano, A., Pacciarelli, D., Pranzo, M.: A branch and bound algorithm for scheduling trains in a railway network. Eur. J. Oper. Res. **183**(2), 643–657 (2007)
4. D'Ariano, A., Pranzo, M., Hansen, I.A.: Conflict resolution and train speed coordination for solving real-time timetable perturbations. IEEE Trans. Intell. Transp. Syst. **8**(2), 208–222 (2007)
5. Kecman, P.: Models for predictive railway traffic management. Ph.d. thesis, Delft University of Technology, Delft (2014)
6. Hansen, I.A., Goverde, R.M.P., Van der Meer, D.J.: Online train delay recognition and running time prediction. In: 13th International IEEE Conference on Intelligent Transportation Systems (ITSC), Madeira, pp. 1783–1788 (2010)
7. Kecman, P., Goverde, R.M.P.: Online data-driven adaptive prediction of train event times. IEEE Trans. Intell. Transp. Syst. **16**(1), 465–474 (2015)
8. Goverde, R.M.P., Hansen, I.A.: Performance indicators for railway timetables. In: IEEE International Conference on Intelligent Rail Transportation (ICIRT), pp. 301–306 (2013)
9. Buker, T., Seybold, B.: Stochastic modelling of delay propagation in large networks. J. Rail Transp. Plan. Manage. **2**(1–2), 34–50 (2012)
10. Medeossi, G., Longo, G., de Fabris, S.: A method for using stochastic blocking times to improve timetable planning. J. Rail Transp. Plan. Manage. **1**(1), 1–13 (2011)
11. Longo, G., Medeossi, G., Nash, A.: Estimating train motion using detailed sensor data. In: Transportation Research Board 91st Annual Meeting, Washington, pp. 1–15 (2012)
12. Meester, L.E., Muns, S.: Stochastic delay propagation in railway networks and phase-type distributions. Transp. Res. Part B Methodol. **41**(2), 218–230 (2007)
13. Carey, M., Kwiecinski, A.: Properties of expected costs and performance measures in stochastic models of scheduled transport. Eur. J. Oper. Res. **83**, 182–199 (1995)
14. Yuan, J., Hansen, I.A.: Optimizing capacity utilization of stations by estimating knock-on train delays. Transp. Res. Part B Methodol. **41**(2), 202–217 (2007)
15. Berger, A., Gebhardt, A., Muller-Hannemann, M., Ostrowski, M.: Stochastic delay prediction in large train networks. In: Caprara, A., Kontogiannis, S. (eds.) 11th Workshop on Algorithmic Approaches for Transportation Modelling, Optimization, and Systems, Dagstuhl, pp. 100–111 (2011)
16. Davydov, B., Chebotarev, V., Kablukova, K.: On stochastic modeling the arrival time deviations. In: RailBeijing2021 (2021, to be published)
17. Chebotarev, V., Davydov, B., Kablukova, K.: Probabilistic model of delay propagation along the train flow. In: Kostogryzov, A. (ed.) Probabilistic Modeling in System Engineering (Chapter 8), pp. 171–193. IntechOpen (2018)
18. Wentzel, E.S.: Probability Theory: Textbook for Universities, Moscow (1999). 576 p. (in Russian)

Check for updates

# Video-Based Background/Foreground Estimation for Vehicle Detection

Zbigniew Czapla(✉)

Faculty of Transport and Aviation Engineering, Silesian University of Technology, Katowice, Poland
zbigniew.czapla@polsl.pl

**Abstract.** The paper describes a new approach to background/foreground estimation in images from a video stream. The proposed method is based on the analysis of small gradients in images. The video stream is recorded with the camera mounted in the measuring station above the road and directed along the road. Input images are converted into binary form and divided into square blocks. For each block the sum of binary values is calculated. Blocks are assigned to the background or to the foreground of an input image. Background/foreground assignment depends on sums of binary values within individual blocks. Vehicle detection is carried out through the analysis of blocks assigned to the foreground. The proposed method of background/foreground estimation is intended for systems of traffic measurements and surveillance.

**Keywords:** Background/foreground estimation · Image gradients · Image partitioning · Vehicle detection

## 1 Introduction

In contemporary road traffic systems image data are used for measurements of traffic parameters and for traffic surveillance. Processing and analysis of acquired image data allow vehicle detection and determination of the number of vehicles traveling on the road in a measurement period. The analysis of images from traffic scenes can be applied for vehicle classification into type category, description of the vehicle trajectory, and also for estimation of the vehicle speed and a distance between vehicles.

In [1], a feature-based tracking method for vehicle detection is presented. The presented method bases on corner features which concern regions with changes of brightness in more than one direction. Corner features are tracked, grouped and traffic parameters are determined with the use of vehicle trajectories. In [2], segmentation is applied for vehicle separation from the background. Processing is performed in several stages and includes subtraction of the current image from the background, threshholding, binarisation and updating of the background. In [3], the method of vehicle tracking and classification is presented. The presented method uses image differencing and background updating. The next method stages contain determination of lane-dividing lines, shadow-elimination and vehicles trucking with the use of Kalman filtering. In [4], traffic monitoring with the use


G. Sierpiński (Ed.): TSTP 2021, LNNS 352, pp. 110–119, 2022.
https://doi.org/10.1007/978-3-030-91156-0_9

the static and dynamic analyses of segmented images is described. The static analysis is utilized for processing of single images and the dynamic analysis involves processing of current and previous images. Applied traffic monitoring provides parameters of traveling vehicles.

In [5], the two-level method of a traffic scene analysis is presented. The presented method contains high-level and low-level modules. The low-level processing modules are destined for extracting objects through the analysis of consecutive frames and the high-level processing module is destined for vehicle tracking with the reasoning technics. In [6] the use of time-spatial images is presented. The time-spatial images correspond to virtual lines on frames of the input video stream. The analysis of the sequence of the time special images allows vehicle detection and classification. In [7], a method of vehicle tracking at intersections is presented. The presented method includes the stochastic algorithm applying division of analysed images into square blocks.

In [8], background modelling with the use of Gaussian distributions is presented. The background model is based on description of pixel values through mixture of Gaussians assigned to individual pixels. In [9] functions of probability density assigned to the background and to the foreground are applied. These statistical representations of the background and the foreground are determined with the use of kernel density techniques. In [10], the selected known methods of background subtraction are concisely described and evaluated. Evaluation is performed with consideration of processing speed, requirements of memory and method accuracy.

In [11], the method of vehicle detection with the use of image data is presented. The presented method is based on the analysis of small gradients in a sequence of input images. Consecutive input images are converted into the binary format and the arithmetic sums of the binary values are calculate within the defined detection field. Vehicle detection is performed through the analysis of the arithmetic sum of the binary values inside the detection field and calculated for consecutive input images.

The proposed method of background/foreground estimation allows assignment of individual image blocks appropriate to the background or to the foreground on the basis of image local gradients. Background/foreground assignment of image blocks is performed for vehicle detection both in the entire analyzed images and in regions defined as detection fields.

## 2 Video Data and Image Binarization

Input video data in the form of a video stream are obtained at a measuring station. The measuring station is equipped with the stationary camera mounted above the road and in the direction along the road. The video stream from the camera consists of video frames. Consecutive frames of the video stream create the sequence of input images. For background/foreground estimation grayscale input images in the bitmap format are used. Each input image is of size $N$ columns by $M$ rows and has the ordinal number in the sequence of input images. The image ordinal number is denoted by $i$ and shows a sequence position of the individual images. During processing input images are successively converted into binary form.

Conversion of input images turns images in the bitmap format into binary images. A single input image is represented by an image matrix denoted by $\mathbf{A}$ and of order $M$ rows by $N$ columns. Elements of the matrix **A** corresponds to pixel values of an input image. A binary matrix denoted by **B** is destined for binary values after conversion and is of the same order as the matrix **A**. Indices of the matrices **A** and **B** denoted by $m$, $n$ are in the range from 1 to $M$ for rows and from 1 to $N$ for columns respectively and corresponds to pixel coordinates of an input image. Matrix row and column indices are supplemented by the third index in parentheses that is equal to the ordinal number in the sequence input image. Conversion of the current input image into binary form consists in processing of elements of an appropriate image matrix successively in rows and columns.

Initially, all elements of the binary matrix are equal to 0. During conversion small gradients are calculated for each processed element of the image matrix in its nearest neighbourhood in a row $G_{r(i)}(m, n)$, in a column $G_{c(i)}(m, n)$ and in diagonal directions up $G_{u(i)}(m, n)$ and down $G_{d(i)}(m, n)$

$$\begin{aligned} G_{r(i)}(m,n) &= a_{(i)m,n} - a_{(i)m,n-1}, \\ G_{c(i)}(m,n) &= a_{(i)m,n} - a_{(i)m-1,n}, \\ G_{u(i)}(m,n) &= a_{(i)m,n} - a_{(i)m-1,n-1}, \\ G_{d(i)}(m,n) &= a_{(i)m,n} - a_{(i)m+1,n-1}. \end{aligned} \tag{1}$$

Magnitudes of determined gradients are compared to the preset threshold gradient value $T_G$. For gradients which the magnitude is greater than the threshold value, elements of the binary matrix **B** which correspond to the beginning and the end of the gradient in the image matrix are set to 1.

$$\begin{aligned} b_{(i)m,n-1} = b_{(i)m,n} = 1 \quad &\text{for} \quad \left| G_{r(i)}(m,n) \right| > T_G, \\ b_{(i)m-1,n} = b_{(i)m,n} = 1 \quad &\text{for} \quad \left| G_{c(i)}(m,n) \right| > T_G, \\ b_{(i)m-1,n-1} = b_{(i)m,n} = 1 \quad &\text{for} \quad \left| G_{u(i)}(m,n) \right| > T_G, \\ b_{(i)m+1,n-1} = b_{(i)m,n} = 1 \quad &\text{for} \quad \left| G_{d(i)}(m,n) \right| > T_G. \end{aligned} \tag{2}$$

After conversion the binary matrix consists of elements which the value is equal to 0 or 1. Elements of 1 corresponds to edges contained in the converted input image, thus these elements in the binary matrix are called the edge points.

In the next stages binary matrices are divided into square blocks of the same size and background/foreground assessment of blocks is carried out on the basis of changeability of the sums of the edge points within individual blocks.

## 3 Partitioning into Blocks

Partitioning of the binary matrix into blocks is carried out after binarization and consists in division of elements in the binary matrix into square blocks of size $d$ by $d$ elements. The position of each block is described by row and column block coordinates denoted by $k$ and $l$ respectively. Ranges of block coordinates are given by

$$\begin{aligned} k = 1 \ldots K,\ K = \left\lfloor \tfrac{M}{d} \right\rfloor, \\ l = 1 \ldots L,\ L = \left\lfloor \tfrac{N}{d} \right\rfloor, \end{aligned} \tag{3}$$

where $K$ and $L$ are the floor of the expressions in brackets. The indices inside individual blocks of binary matrices belong to the intervals

$$\begin{aligned} m &= (k-1)d + 1 \ldots kd \ , \\ n &= (l-1)d + 1 \ldots ld \ . \end{aligned} \tag{4}$$

For each block of the binary matrix the sum of the edge points is calculated. The sums of the edge points within blocks are elements in the image block matrix denoted by **S**. The image block matrix is of order $K$ rows by $L$ columns. Elements of the image block matrix are calculated as the sums of edge points within blocks of elements in the binary matrix as follows:

$$s_{(i)k,l} = \sum_{m=(k-1)d+1}^{kd} \ \sum_{n=(l-1)d+1}^{ld} b_{(i)m,n} : b_{(i)m,n} = 1 \ . \tag{5}$$

Elements of the image block matrix are destined for assignment of individual blocks to the image foreground or background.

The background block matrix, denoted by **R,** of the same order $K$ by $L$ as the image block matrix, is destined for the sums of edge points of binary matrix blocks assigned to the background. Elements of the background block matrix should be initialized before background/foreground block assignment is applied for the first analysed input image. Initialization of elements of the background block matrix is carried out through calculation of average block sums for the preset number of consecutive images preceding the first analysed input image. The number of the considered images preceding the first analysed input image is denoted by $I_{\text{ini}}$. The initial values of elements of the background block matrix are calculated as follows:

$$r_{(i)k,l} = \frac{1}{I_{ini}} \sum_{i=-I_{ini}+1}^{0} \ \sum_{m=(k-1)d+1}^{kd} \ \sum_{n=(l-1)d+1}^{ld} b_{(i)m,n} : b_{(i)m,n} = 1 \tag{6}$$

The number of images considered for initialisation of the background block matrix should be of a few hundreds of greater.

Except the image block matrix and the background block matrix, a pixel background block matrix is defined. The pixel background block matrix of order $K$ by $L$, denoted by **X,** is destined for the sums of elements in the image matrix within individual blocks assigned to the background. Similarly to elements of the background block matrix, elements of the pixel background block matrix need initialization. The initial values of elements in the pixel block matrix are calculated as follows:

$$x_{(i)k,l} = \frac{1}{I_{ini}} \sum_{i=-I_{ini}+1}^{0} \ \sum_{m=(k-1)d+1}^{kd} \ \sum_{n=(l-1)d+1}^{ld} a_{(i)m,n} \ . \tag{7}$$

The number of images considered for initialisation of the pixel block matrix is the same as for initialisation of the background block matrix.

## 4 Background/Foreground Estimation

Background/foreground estimation of regions in the current input image is carried out through assignment of blocks of elements in the binary matrix to the background or to the foreground. The pointer block matrix, denoted by **P** and of the same order $K$ by $L$ as the image block matrix, is destined for background/foreground pointers. These pointers indicate the blocks of the current image matrix as belonging to the background or to the foreground. Elements of the pointer block matrix are determined through comparison of a difference between appropriate blocks of the input block matrix and the background block matrix to the preset block threshold value, denoted by $T_B$, according to the expression

$$p_{(i)\,k,l} = \begin{cases} 0 & \text{for} \quad \left| s_{(i)\,k,l} - r_{(i)\,k,l} \right| \leq T_B \ , \\ 1 & \text{for} \quad \left| s_{(i)\,k,l} - r_{(i)\,k,l} \right| > T_B \ . \end{cases} \tag{8}$$

Elements of the pointer block matrix equal to 0 mean blocks assigned to the background and elements equal to 1 blocks assigned to the foreground.

Foreground attributed image blocks represent moving objects that occur in input images. Because background/foreground block assignment is carried out on the basis of the sum of the edge points within blocks, object blocks contain the small number of the edge points can be incorrectly assigned to the background instead of to the foreground. To improve efficiency of background/foreground block assignment, supplementation of foreground blocks can be performed and then the foreground blocks are supplemented with the adjoined nearest neighbouring background blocks. For supplementation of foreground blocks, digital filtering of elements in the current pointer block matrix is applied. Supplementation of the current pointer block matrix is performed in two stages succeeding one by one. In the first stage elements of the current pointer block matrix are processed in pairs of rows in the direction of ascending column index. In the current pointer block matrix, elements are supplemented according to the expression

$$p_{(i)\,k,l+1} = p_{(i)\,k+1,l} = 1 \quad \text{for} \quad p_{(i)\,k,l} = p_{(i)\,k+1,l+1} = 1 \ . \tag{9}$$

In the second stage elements of the current pointer block matrix are also processed in pairs of rows, but in the direction of descending column index and then supplementation is given by the expression

$$p_{(i)\,k,l} = p_{(i)\,k+1,l+1} = 1 \quad \text{for} \quad p_{(i)\,k,l+1} = p_{(i)\,k+1,l} = 1 \ . \tag{10}$$

Background/foreground assignment can by detail through the additionally analysis consisting in comparison elements of the pixel background block matrix to the block sums calculated for blocks of the current image matrix. The block sums of the current image matrix are elements in the detail block matrix of order $K$ by $L$ denoted by **D**. Elements of the current detail block matrix are calculated as follows:

$$d_{(i)k,l} = \sum_{m=(k-1)d+1}^{kd} \ \sum_{n=(l-1)d+1}^{ld} a_{(i)m,n} \ . \tag{11}$$

During the additionally analysis, the differences between corresponding elements of the detail block matrix and the pixel background block matrix are calculated and compared to the preset pixel block threshold value. The result of comparison allows block assessment to the background or to the foreground. The additionally analysis of the current detail block matrix can be limited to the nearest neighborhood of blocks assigned to the foreground.

## 5 Vehicle Detection

Input images in the bitmap format are processed into binary form and after processing divided into blocks. For each block the number of the edge points is calculated within the block. Blocks are appropriate assigned to the background or the foreground of the input image. Analysis of foreground blocks allow detection of vehicles appear and disappear in the input video stream.

Vehicles occur the input video stream can be detected through analysis of elements in consecutive pointer block matrices. Background/foreground approach allows enhancement of vehicle detection efficiency in image areas defined as detection fields. Moreover analysis of position of foreground blocks enables labelling and tracking detected vehicles in the sequence of input images.

After assignment of blocks of the current image appropriate to the background or to the foreground, elements of the current background block matrix are updated. Updating of the current background block matrix consists in calculation of a new value of blocks that belong to the background of the current input image. For each block assigned to the background, the new value is calculated. The new block values are calculated as the average of elements from the current block image matrix and the preset number of the previous consecutive background block matrices. The preset number of considered previous background block matrices is denoted by $I_{\text{pre}}$. The new values of elements in the successive background block matrix is calculated according to the expression

$$r_{(i+1)\,k,l} = \frac{r_{(i)\,k,l} \cdot I_{pre} + s_{(i)\,k,l}}{I_{pre} + 1} \quad \text{for} \quad p_{(i)\,k,l} = 0\,. \tag{12}$$

The new values are calculated for elements of the current background block matrix corresponding to the background of the current input image. The elements of the current background block matrix corresponding to the foreground of the current input image remain unchanged

$$r_{(i+1)\,k,l} \;=\; r_{(i)\,k,l} \quad \text{for} \quad p_{(i)\,k,l} = 1\,. \tag{13}$$

Similarly to updating of elements in the current background block matrix, updating of elements in the current pixel block matrix is carried out after vehicle detection. Updating of elements in the current pixel block matrix consists in calculation of the new values of these elements in the current pixel block matrix that differ from corresponding elements in the detail block matrix less than the preset threshold value denoted by $T_x$. The new values of elements in the current pixel block matrix are calculated according to the expression

$$x_{(i+1)\,k,l} = \frac{x_{(i)\,k,l} \cdot I_{pre} + d_{(i)\,k,l}}{I_{pre} + 1} \quad \text{for } \left| x_{(i)\,k,l} - d_{(i)\,k,l} \right| < T_x \,. \tag{14}$$

Updating of elements in the current background block matrix and elements in the current pixel block matrix allows reducing influence various factors on measurement results. Consistently updating of background features limits impacts of changeably weather conditions, scene illumination, time of the day and also possible changeability of input image quality.

## 6 Experimental Results

Measurements have been taken at the measuring station. The video stream has been obtained in good weather conditions. The frame rate of the obtained video stream is of 30 frames per second (fps). The sequence of input images consists of grayscale images corresponding to the sequence of frames in the video stream. Each input image has intensity resolution 8 bits per pixel (bpp) and is of size 384 by 384 pixels.

Conversion of input images into binary form is carried out with the gradient threshold value preset $T_G = 8$. Elements of the binary matrix are divided into blocks of the same size 32 by 32 pixels. In the beginning initialization of the background block matrix is performed. After initialization of the background block matrix, images from the sequence of input images are processed. Background/foreground assignment is carried out with the block threshold value $T_B$ preset to 0.3 of the maximum number of the edge points within a single block. After background-foreground block assignment, digital filtering is carried out and the background is updated.

The example of images after background/foreground estimation corresponding to a single input image shows Fig. 1. The figure contains an input image, a binary image and a background/foreground image.

**Fig. 1.** Input, binary and background/foreground images

The input image and the binary image are represented by the image matrix and the binary matrix respectively. The background/foreground image is combination of the input image and the binary image depending on the pointer block matrix. In the background/foreground image, the background is composed of blocks of elements from the binary matrix and the foreground is highlighted in blocks of elements from the image matrix. In the binary image black dots denote the edge points.

Although background-foreground estimation based on the analysis of the edge points is intended for vehicle detection inside a detection field defined in input images, background-foreground assignment of image blocks can be applied for vehicle labelling and tracking. The examples of images after background/foreground estimation applied to succeeding in 1/6 s apart input images show Fig. 2–7. Figures include background/foreground images obtained with digital filtering and without the analysis of adjusting blocks. Each figure consists of an input image, a binary image and a background/foreground image. Images presented in Fig. 2 correspond to the image with the ordinal number $j$ in the sequence of the input images.

**Fig. 2.** Input, binary and background/foreground images ($i = j$)

**Fig. 3.** Input, binary and background/foreground images ($i = j + 5$)

**Fig. 4.** Input, binary and background/foreground images ($i = j + 10$)

After background/foreground estimation the moving vehicle presented in the figures is correctly distinguished from the background at various distances from the camera. Background/foreground estimation based on the analysis of the edge points can be

**Fig. 5.** Input, binary and background/foreground images ($i = j + 15$)

**Fig. 6.** Input, binary and background/foreground images ($i = j + 20$)

**Fig. 7.** Input, binary and background/foreground images ($i = j + 25$)

used for vehicle tracking. Application of background/foreground estimation for vehicle tracking requires additionally determination of vehicle positions and labeling of vehicles detected in individual images.

## 7 Conclusions

Background/foreground estimation based on the analysis of the edge points uses conversion of input images in the bitmap format into binary form. During conversion into binary form, calculation of small gradients in input images is applied. Binary images are divided into square blocks and then blocks are appropriate assigned to the background or to the foreground of current input images. After processing, vehicles are represented by adjacent foreground blocks corresponding to the content of input images.

The proposed method of background/foreground estimation allows detection of moving vehicles in the traffic scene recorded in a video stream. The method is intended for efficiency enhancement of vehicle detection within detection fields defined for the traffic scene. Moreover, the presented method of background/foreground estimation can be also applied for vehicle recognition and tracking in the entire area of input images.

Methods for background/foreground estimation are usually complicated and demanding high computing power. The proposed method of background/foreground estimation is computationally simple and fast but not as precise as methods based on the analysis of individual pixels. The presented method is intended for traffic measurements and surveillance and can be suitable for real time traffic systems.

## References

1. Coifman, B., Beymer, D., McLauchlan, P., Malik, J.: A real-time vision system for vehicle tracking and traffic surveillance. Transp. Res. Part C **6**, 271–288 (1998)
2. Gupte, S., Masoud, O., Martin, R.F.K., Papanikolopoulos, N.P.: Detection and classification of vehicles. IEEE Trans. Intell. Transp. Syst. **3**(1), 37–47 (2002)
3. Hsieh, J., Yu, S.H., Chen, Y.S., Hu, W.F.: Automatic traffic surveillance system for vehicle tracking and classification. IEEE Trans. Intell. Transp. Syst. **7**(2), 175–187 (2006)
4. Fernandez-Caballero, A., Gomez, F.J., Lopez-Lopez, J.: Road traffic monitoring by knowledge-driven static and dynamic image analysis. Expert Syst. Appl. **35**, 701–719 (2008)
5. Cucchiara, R., Piccardi, M., Mello, P.: Image analysis and rule-based reasoning for a traffic monitoring system. IEEE Trans. Intell. Trans. Syst. **1**(2), 119–130 (2000)
6. Mithun, N.C., Rashid, N.U., Rahman, S.M.M.: Detection and classification of vehicles from video using multiple time-spatial images. IEEE Trans. Intell. Transp. Syst. **13**(3), 1215–1225 (2012)
7. Kamijo, S., Matsushita, Y., Ikeuchi, K., Sakauchi, M.: Traffic monitoring and accident detection at intersections. IEEE Trans. Intell. Transp. Syst. **1**(2), 108–118 (2000)
8. Stauffer, C., Grimson, W.E.L.: Adaptive background mixture model for real-time tracking. In: Proceedings IEEE Computer Society Conference on Computer Vision and Pattern Recognition, vol. 2, pp. 246–252 (1999)
9. Elgammal, A., Duraiswami, R., Harwood, D., Davis, L.S.: Background and foreground modelling using nonparametric kernel density estimation for visual surveillance. Proc. IEEE **90**(7), 1151–1163 (2002)
10. Piccardi, M.: Background subtractions techniques: a review. In: IEEE International Conference on Systems, Man and Cybernetics, vol. 4, pp. 3099–3104 (2004)
11. Czapla, Z.: Point image representation for efficient detection of vehicles. In: Burduk, R., Jackowski, K., Kurzyński, M., Woźniak, M., Żołnierek, A. (eds.) Proceedings of the 9th International Conference on Computer Recognition Systems CORES 2015. AISC, vol. 403, pp. 691–700. Springer, Cham (2016). https://doi.org/10.1007/978-3-319-26227-7_65

# Improving the Efficiency of Ecological Friendly Solutions in Transport Systems

# The Areas of Impact of the Circular Economy in Passenger Air Transport

Marcin Rutkowski(✉)

Faculty of Economics, The University of Economics, 1 Maja 50, 40-287 Katowice, Poland
marcin.rutkowski@edu.uekat.pl

**Abstract.** The article is a theoretical consideration of identifiable areas of the circular economy in passenger air transport and its determinants, taking into account its ecological character. The article describes forms of aircraft financing that allow the operation of a given aircraft by several carriers in the long term of the aircraft's airworthiness. It also uses a case study to present the advantages of recycling and upcycling aircraft and presents data on $CO_2$ emissions, which may depend on the class of travel in passenger air transport. The closed-loop economy can be an interesting model for the management of aircraft, their components within the recycling and upcycling processes because by its pro-environmental character it can indicate the optimal lifetime of a given good and point to the possibility of using this good in a different, utilitarian way for humans. Air passenger transport is an example of a market where circular economy areas can be identified. The process of financing aircraft fleets is also an area of influence of the circular economy Thanks to recycling and upcycling processes, used components of passenger aircraft construction can be recycled and used both in the aviation industry and in other branches of the economy. The circular economy model can be an effective solution for resource management in passenger air transport.

**Keywords:** Circular economy · Passenger air transport · Recycling · Upcycling · Lease

## 1 Introduction

The circular economy is an extremely important, extensive and new research topic. This is evidenced by the fact that the beginning of the scientific debate on the circular economy is in the 1990s of the twentieth century [1], and the concept of the circular economy itself is expected to gradually displace the linear economy as the dominant model of managing certain goods [2]. The idea of circular economy was based on the observation of natural ecosystems, in which various goods provided by nature, become needed by subsequent entities in the next cycles of their existence. In this way, a kind of closed-cycle is created, in which there should be no waste of a particular good and production of unnecessary waste. The principle of the circular economy seems to be extremely simple, as it is based on the concept of the circulation of specific goods at a specific time. The circular economy is based mainly on respect for the environment by extending the life cycle of a given good more efficiently and effectively.


G. Sierpiński (Ed.): TSTP 2021, LNNS 352, pp. 123–132, 2022.
https://doi.org/10.1007/978-3-030-91156-0_10

It can be assumed that the development of the concept of the circular economy was the result of significant reflection of countries on the direction, quality and pace of development of the world, perceived as the effect of the interaction of the economies of individual countries.

While it can be argued that the circular economy is a new model that organizes the management of certain goods, the very idea of a closed-loop, the idea of circulation, has its historical foundations in many research fields. In science and natural sciences, the concept of circulation, as well as the idea of circulation, can be seen in the microscopic image, studying the structure of atoms, describing the movement of electrons, analyzing the circulation of elements in nature. In physics, energy is defined as the state of a system, of matter, and according to the principle of conservation of energy, it is not possible to create or destroy it, but only to change it from one form into another.

The idea of the circulation of specific goods can also be found in cultural anthropology. An example is the observation of rituals and certain behaviors of social groups. B. Malinowski described the Kula ritual, which took place among the tribes inhabiting the Triobrand Islands [3]. The idea of the ritual was to circulate certain goods in the form of jewelry made of shells with symbolic and ritual significance among the tribes of New Guinea. Concerning Egyptian culture, the idea of circulation can be seen in the ritual of burial of rulers along with the placement of objects in tombs, mummification of corpses, as well as in the concept of reincarnation.

This paper aims to identify areas of circular economy in passenger air transport with the indication of pro-ecological solutions in the air transport branch.

The article consists of four chapters. The introduction presents the characteristics of circular economy from the perspective of differentiated research areas and refers to historical events that may provide evidence that the circulation of certain goods has previously occurred in given cultural circles. The second chapter describes the circular economy by analyzing the literature on the subject along with its determinants. The third chapter identifies areas of circular economy impact on passenger air transport. Leasing, recycling and upcycling as well as $CO_2$ emissions are indicated as processes that can be described within the framework of the circular economy. The final chapter summarizes the areas of impact of the circular economy on passenger air transport.

The article is a theoretical consideration of identifiable areas of the circular economy in passenger air transport and its determinants, taking into account its ecological character. A case study was used to identify areas of circular economy impact concerning examples of individual stakeholder activity in passenger aviation recycling and upcycling, and carbon emissions figures for a given class of passenger air travel were used.

## 2 Circular Economy and Its Determinants

When considering the subject of the closed-loop economy, one encounters various descriptions of it and attempts to define areas of impact. In the literature, one can point to the publication of K. Boulding, who in his essay [4] considers the Earth as a closed ecosystem and describes the functioning of closed and open systems and the relationships between them through the prism of economics. A. Murray points out that the closed-loop economy is a counterbalance to the dominant linear economy models [5].

One current and still valid definition describes the circular economy as an economy that aims to remanufacture, reproduce goods or molecules, helping to maintain the highest level of utility of the newly created components [6].

It can be noted that the principles of closed-loop economics, significantly differ from the commonly known linear model of an economy, in which the economy is described in terms of activities: production, use and disposal of a specific good. It is also interesting to note that the circular economy was implemented through the practices of business people, politicians, large companies, foundations and associations. It was a man through his observations that transformed the economy which proved to be extremely important in different areas of life. The circular economy has been built under the slogan of energy efficiency and maximum use of waste, which are the defects of consumption and manufacturing processes.

The circular economy has a significant impact on the ideas of sustainable development. In addition to its core assumptions, the circular economy model, also indicates more and more other related solutions. One of them is striving to stop buying products and services by consumers, in favour of switching to the form of renting or temporary use of a given good. In this aspect, circular economy fits into the sharing economy model. As an example of such a model, we can point to such initiatives as carsharing and bike-sharing. It is possible that through such activities, consumers will contribute to reducing their negative impact on the environment. It can also be assumed that future generations will drastically change the way they think and approach consumption and production.

Pursuing a circular economy can have tangible benefits. The circular economy can be defined as a model of economy aiming at rational, effective and efficient use of given resources in the form of specific raw materials, goods, assets while minimizing the negative impact of these resources on the environment. The determinants by which the circular economy is described include innovation, protection and respect for the environment, competitiveness concerning the linear economy, sustainability of products and their regenerative nature, security of raw material supply, savings and increase in the quality of life as well as stable economic growth based on sustainable development [7]. It is also worth emphasizing the fact that closed-loop economics, corresponds with ecology, sozology and other fields of science that supports the economic and regenerative way of managing certain goods.

It seems that a necessary condition for the effective and efficient operation of the circular economy model is close cooperation between the public, private and non-profit sectors, both in the micro-and macroeconomic dimensions. Joint efforts to develop a circular economy should be based mainly on making rational and responsible decisions that will result in numerous benefits for the economies of individual countries in the future. How to do it in practice? The answer to this question is not unambiguous and simple, because the superiority of man over nature is becoming more and more visible and destructive, and the man himself often becomes a part of one of the interest groups that do not always have to strive for the realization of circular economy assumptions.

Air passenger transport is an example of a transport branch where it is possible to identify the areas of influence of the circular economy taking into account its pro-ecological character, i.e. favoring the preservation of balance in nature. The pro-ecological dimension of circular economy in the passenger air transport may be expressed in: the period of aircraft exploitation by airlines, forms of aircraft acquisition by carriers, as well as in the way of processing end-of-life aircraft into new, useful goods.

## 3 Leasing, Recycling and Upcycling as Elements of the Circular Economy in Air Passenger Transport

The necessity of implementing circular economy in passenger air transport is indicated by ICAO. The International Civil Aviation Organization refers in this aspect to the three principles of the Ellen McArthur Foundation [8]. The implementation of a closed loop economy in passenger air transport addresses three overarching principles:

- waste is food - redefining the purpose of end-of-life products can extend their life and reduce the environmental impact of new product manufacturing. Within a closed loop, proper maintenance, reuse, refurbishment and recycling can extend the life cycle of products. These products are no longer considered as waste, but as necessary inputs for manufacturers and service providers and can be reintegrated into the lifecycle data.
- using renewable resources - by increasing the use of renewable or waste-derived resources and energy, a circular economy model could create new types of jobs and reduce environmental impacts, including carbon emissions.
- building resilience through diversity - to achieve reductions in the use of primary materials and the generation of waste, supply chains must be designed to refocus products from one production process to another. Therefore, designing a circular economy model requires bringing together diverse companies and stakeholders with different functions within a circular economy system.

The closed-loop economy is believed to be an emerging concept in the civil aviation sector, and while its application is still not widespread, the use of closed-loop economy concepts can already provide valuable environmental, social and economic opportunities. The closed-loop economy has the potential to transform the entire supply chain, from product design to end-of-life management.

It can be assumed that passenger air transport is already using some of the concepts associated with the circular economy. It mainly focuses on three elements of aviation: aircraft, flights and airports. Manufacturers, airlines and airports are already involved and more are expected in the future. There are many opportunities, such as reuse and recycling of end-of-life (EoL) aviation materials, reduction and valorization of flight and ground infrastructure waste (food, packaging, municipal solid waste, construction and demolition debris), reduction of energy consumption, water consumption, and raw material consumption.

An example of the impact of the circular economy is passenger air transport. In the dimension of the circular economy, passenger air transport as a component of a department of the national economy can be described on several levels of organization and functioning. Regarding the question of the financing of passenger aircraft as a means of transport by airlines, it is possible to point to various types of lease as a form of acquisition of both new and used passenger aircraft. The acquisition of an aircraft by an airline is a complex process.

The lease agreement is a form of aircraft financing by airlines can be an example of how a closed-loop economy works. Expensive assets, which can be considered aircraft, can be exchanged within air carriers depending on the demand for air transport services. A temporary rental eliminates the need for the airline to purchase the aircraft.

Airlines can purchase an aircraft for cash, directly from the aircraft manufacturer, or sign a lease agreement with a lessor, which is the more preferred form for carriers to acquire extremely expensive assets [9]. When it comes to lease contracts, there are three basic types of lease contracts. The first type of lease agreement is the dry lease, which comes in the form of both operating and finance leases. Under this type of agreement, an airline acquires an aircraft and uses its flying and cabin crew resources. Typically, this type of agreement is signed for a period of two to several years. At the end of the lease, depending on whether it is an operating or a finance lease, the aircraft is either returned to the lessor or bought back by the airline and becomes a book asset. Dry leasing is part of an airline's long-term growth strategy.

Another type of passenger aircraft lease arrangement is called damp lease, where the lessor is required to provide the aircraft with pilots, is also responsible for the aircraft's fitness for flight operations, and is responsible for training the cabin crew to offer onboard service and maintenance. The airline, on the other hand, is required to ensure the presence of cabin crew. The third type of lease is the so-called wet lease, under which the lessor is required to provide the lessee with an aircraft that is fit for flight operations, together with flight and cabin crew, and to insure the aircraft. Often wet leasing is also referred to as ACMI- Aircraft, Crew, Maintenance, Insurance. Typically, these contracts are signed by airlines for 24 months and are an example of an airline's short term development strategy. Wet leasing is a form of financing aircraft, mainly in the case of a seasonal increase in demand for transport services on a given route or in the case of grounding part of the carrier's fleet. A common practice, which is a realization of the circular economy, is the conversion of previously operated passenger aircraft into aircraft carrying cargo.

Passenger aircraft lease agreements apply to both new and used passenger aircraft. The useful life of a passenger aircraft may exceed thirty years. During this period, the passenger aircraft can be operated by many carriers, with safety in the technical dimension, which determines the fitness of the aircraft for flight operations.

Another aspect in the operation of passenger air transport concerning the circular economy is the management of $CO_2$ emissions, a gas that contributes to the carbon footprint as one of the effects of aircraft operation. The minimization of the carbon footprint by air carriers is part of the green development policy of passenger air transport, which is one of the domains of the circular economy. Airlines not only cooperate with aircraft manufacturers to reduce $CO_2$ emissions but also implement pro-environmental actions to offset $CO_2$ emissions by implementing programs that support the restoration or regeneration of specific ecosystems. An example of such actions can be the activity of Australian airline Qantas. As part of $CO_2$ offsetting projects, the airline actively participates in helping indigenous tribes to prevent fires and supports the regeneration of the Great Barrier Reef [10].

An interesting phenomenon is the variable number of kilograms of $CO_2$ emissions depending on the class of travel. Travelling on a selected route, a passenger in premium class will have a different value of kilograms of $CO_2$ emissions than a passenger travelling on the same aircraft in economy class. This is due, among other things, to the different weight of the seats, the number of in-flight products offered and the increased number of baggage allowances per class of travel.

As part of the actions taken by the airlines in the implementation of the circular economy assumptions, we can also point to the gradual phasing out of the use of cutlery and objects serving food and beverages as part of the available onboard service. An important measure is also the possibility to use electronic boarding passes, without the need to have paper versions.

Another area of influence of the circular economy on passenger air transport is the recycling and upcycling of aircraft. Recycling concerning passenger air transport can be defined as the recovery of individual aircraft components whereby waste is reprocessed into products, materials or substances for the original purpose or other purposes. The recycling process can also be explained as the reuse of waste that becomes a new, completely different product. Recycling in a broader sense is also about taking care of the environment by reducing the continuous use of natural resources and minimizing waste.

In 2015, Air France-KLM Group decided to implant circular economy solutions [12], which it chose to base on four pillars: redesign: by redesigning catering services to properly segregate waste - reduce: by reducing the weight of food packaging and moving instructions from paper to digital; - reuse: by reusing seats and on-board entertainment systems in other systems; and - recycling: by recycling reusable equipment, including trays, drawers, blankets and carts, etc.

Upcycling is a process similar to recycling. It is also based on the secondary processing of waste. However, the products that are created using it have a higher value than the raw materials used. Used and seemingly unnecessary goods go into another, re-circulation, in which they gain new specified functions, and also take a given form.

Currently, one can notice the tendency of a significant increase in the rank of certain branches of transport in a given country, which may be a consequence of the development trends of selected economies. It can also be assumed that an increasing influence on the development of countries plays passenger air transport. Consequently, the demand for air transport services translates into the development of the passenger air market (Table 1).

Considering passenger aircraft in terms of service life, it can be concluded that their characteristic feature is long service life. Passenger aircraft can be used by airlines for more than thirty years. It is worth noting that the service life of a particular type of passenger aircraft may depend on the number of flight operations cycles performed, operating conditions, or specific aviation events. Paradoxically, the long service cycle of a given aircraft is one of the reasons for the lack of appearance of an idea for the later use of worn-out aircraft. However, when a passenger aircraft can no longer be operated it is placed in a storage yard, where it awaits demolition and processing of the building elements into new items. Many reasons can also be identified for the end of the use of a particular aircraft. They usually occur when the costs of repairing individual

**Table 1.** Hypothetical number of kg of $CO_2$ emissions per passenger for selected routes

| Route | Travel class | Number of kg of $CO_2$ per passenger | Aircraft type |
|---|---|---|---|
| Warsaw-Frankfur-Warsaw | Economic | 225,2 | Airbus A319, A321, Boeing 737, Embraer 195 |
| Warsaw-Frankfur-Warsaw | Premium | 225,2 | Airbus A319, A321, Boeing 737, Embraer 195 |
| Warsaw-London-Warsaw | Economic | 275,0 | Airbus A319, A321, Boeing 737, Embraer 195 |
| Warsaw-London-Warsaw | Premium | 275,0 | Airbus A319, A321, Boeing 737, Embraer 195 |
| Warsaw-Toronto-Warsaw | Economic | 719,7 | Boeing 787-8, 787-9 |
| Warsaw-Toronto-Warsaw | Premium | 1439,5 | Boeing 787-8, 787-9 |
| Warsaw-Tokyo-Warsaw | Economic | 645,0 | Boeing 787-8, 787-9 |
| Warsaw-Tokyo-Warsaw | Premium | 1289,9 | Boeing 787-8, 787-9 |
| Warsaw-Beijing-Warsaw | Economic | 564,8 | Boeing 787-8, 787-9 |
| Warsaw-Beijing-Warsaw | Premium | 1129,6 | Boeing 787-8, 787-9 |

Source: Own elaboration based on [11].

components, acquiring spare parts systematically increase, and technological changes become too costly, or when the time of safe operation has come to an end.

The ageing process of aircraft becomes the primary reason for their subsequent demolition. At a certain stage, technological upgrades or repair attempts become unprofitable for airlines. In such circumstances, lessors or airlines usually decide to disassemble functional systems and components and try to sell them. In many cases, the value of individual components is higher than the value of the whole aircraft. However, not all parts can be reused, and even worse, they can become an environmental hazard and nuisance if not properly processed into a form that does not adversely affect the environment. In most cases, however, end-of-life aircraft are dismantled and then sold to other airlines. As a last resort, they are placed in warehouses or deserts that serve as storage yards.

A significant problem arises when a manufacturer decides to end the production of a given model of a passenger plane and individual airlines withdraw a particular model from their fleet. An example of this is the Boeing Company's decision to end production of the Boeing 747 and the withdrawal of this model from service by many airlines worldwide. The Boeing 747 is composed of nearly 66 tons of aluminium, nearly 250 km of cables, and 8 km of tubes, totaling nearly 6 million components [13]. One of the most valuable components of the Boeing 747 is its engine, which can be sold for nearly $3 million.

It turns out that once an aircraft is taken out of service, it is possible to propose reusing its components.

As an example of the implementation of aircraft, recycling can be considered the activity of the German company Aviationtag [14], which uses fuselages from scrapped aircraft and transforms them into the form of keychains. The miniature souvenir features information in the form of an individual aircraft number and flight data. As it turns out, fuselage parts can also be used as wall clocks, and appropriately processed screws are used as cufflinks [15]. Engine fans are also used as table bases and as interior design elements.

Boardbar Design GmbH, a company, has based its business on converting on-board riding buffets into bathroom cabinets, wine coolers and document containers [16]. In addition to aircraft parts, the outfits of the flight and cabin crew are also upcycled. In the next life cycle, they are used as pillows, bags and covers. An interesting option is to use used aircraft for training purposes. The most common use is for cockpits, where pilots can hone their skills. Sometimes, the plane is also used as an element of training ground infrastructure for the uniformed services. In many cases, end-of-life airplanes are repurchased by entrepreneurs and used in a completely different way. They are converted into hotels, restaurants or become collector's items.

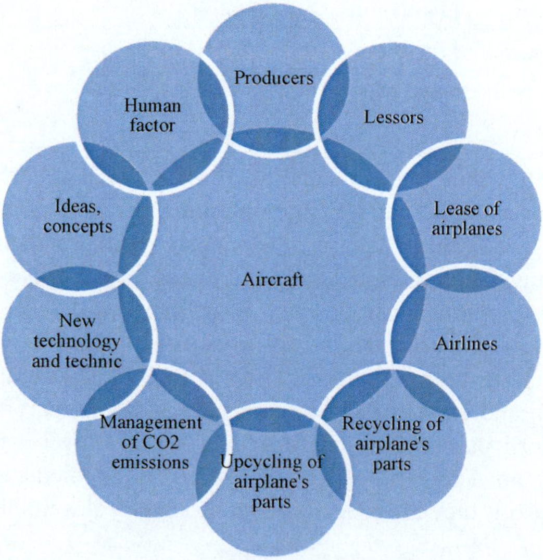


**Fig. 1.** The greening of the circular economy in passenger air transport

Leading aircraft manufacturers such as Boeing Company and Airbus are also implementing programs to make air transport environmentally friendly. European manufacturer Airbus has developed the PAMELA project (Process for Advanced Management of End-of-Life Aircraft) [17], which demonstrates the recyclability of up to 85% of the plane's components, a significant improvement over the previous rate of 60%. This action was taken in response to a large number of aircraft that will be retired over the next few years and its environmental and economic impact. The Boeing Company proposed the AFRA program (Aircraft Fleet Recycling Association) [18] to establish a new standard for environmentally responsible end-of-life aircraft management. Founded in 2006, AFRA is a global membership-based collaboration to improve industry efficiency and commercial value for end-use aircraft. AFRA represents companies from around the world and across the supply chain - from manufacturers to materials recyclers. Through the collective experience of its members, AFRA's Best Management Practices (BMP) guide has significantly improved the environmental and sustainability management of end-of-life aircraft. AFRA's Best Practices for the Management of Used Aircraft Parts and Assemblies and the Recycling of Aircraft Materials (BMPs) is a document that provides a set of recommendations for best practices for managing parts that are removed from an aircraft, engine, or other asset when the asset is dismantled at the end of its useful life and for recycling parts and materials that are recovered from an aircraft, engine, or other asset when the asset is recycled at the end of its useful life [19].

## 4 Summary

The concept of circular economy is an interesting model of managing certain resources. The very idea of the circulation of goods has its roots in historical events and can also be observed in the research field of many scientific disciplines. The assumptions of the circular economy can also be identified in passenger air transport. The operation of the circular economy model in civil aviation manifests itself in the purchase and sale of passenger aircraft, both on the primary and secondary market. The processes of passenger aircraft exchange between air carriers are fostered by various forms of lease agreements as tools through which it is possible to finance expensive assets. Another example confirming the functioning of the circular economy is the aspect of $CO_2$ and other greenhouse gas emissions, which are one of the external effects of passenger air transport. The airlines try to mitigate the negative effects of harmful gas emissions by more efficient management of carbon footprint, as well as by compensation measures. With the help of recycling and upcycling processes, individual components of aircraft construction and equipment are integrated into subsequent use cycles.

The implementation of a circular economy in passenger air transport can confirm that the aviation sector wants to make a significant contribution to a sustainable development-oriented towards the economical management of certain goods. This implementation can take place in cooperation with economic and technical achievements and technologies. The key remains the type of propulsion system for passenger air transport and the materials and processes for building passenger aircraft.

The circular economy model can be an effective solution for resource management in passenger air transport. Another important issue is the search for cost-effective propulsion systems and designing innovative materials for aircraft construction, as actions in line with the assumptions of the closed-circle economy.

## References

1. Pieńkowski, D., Kośmicki, E.: Funkcja produkcji gospodarki zamkniętego obiegu. W: Ekonomia i środowisko, nr 2 (2016)
2. Qiao, F., Qiao, N.: Circular economy: an ethical and sustainable economic development model. In: Prakseologia nr 154 (2013)
3. Malinowski, B.: Argonauci Zachodniego Pacyfiku: relacje o poczynaniach i przygodach krajowców z Nowej Gwinei, B. Olszewska-Dyoniziak (tłum.) i inni, PWN, Warszawa (2005)
4. Boulding, K.E.: The economics of the coming spaceship earth. In: Jarret, H. (eds.) Environment Quality in a Growing Economy. Essays form the Sixth RFF Forum, Baltimore, pp. 3–14, Baltimore (1996)
5. Murray, A., Skene, K., Haynes, K.: The circular economy: an interdisciplinary exploration of the concept and application in a global context. J. Bus. Ethics **3**, 369 (2015)
6. Cullen, J.M.: Circural economy: theoretical benchmark or perpetual motion machine? J. Ind. Ecol. **21**, 483 (2017)
7. Geisendorf, S., Pietrulla, F.: The circular economy and circular economic concepts-a literature analysis and redefinition. Thunderbird Int. Bus. Rev. **60**, 771 (2018)
8. ICAO Introduction to Circular Economy (2021). https://www.icao.int/environmental-protection/Documents/EnvironmentalReports/2019/ENVReport2019_pg275-278.pdf. Version from 02 March 2021
9. Bourjade, S., Huc, R., Muller-Vibe, C.: Leasing and profitability: empirical evidence from the airlines industry. Transp. Res. Part A **97**, 30 (2017)
10. Qantas (2021). https://www.qantas.com/au/en.html. Version from 02 March 2021
11. ICAO Carbon Emission Calculator (2020). https://www.icao.int/environmental-protection/CarbonOffset/Pages/default.aspx. Version from 15 December 2020
12. Sustainability Report 2019 (2021). https://sustainabilityreport2019.airfranceklm.com/waste-and-circular-economy/. Version from 03 March 2021
13. Boeing 747 (2021). https://www.boeing.com/commercial/747/. Version from 02 March 2021
14. Aviationtag (2021). https://www.aviationtag.com. Version from 03 March 2021
15. Plane Industries (2021). https://www.planeindustries.com. Version form 04 March 2021
16. Bordbar (2021). https://www.bordbar.de. Version from 04 March 2021
17. Process for Advanced Management of End of Life of Aircraft (2021). https://ec.europa.eu/environment/life/project/Projects/index.cfm?fuseaction=search.dspPage&n_proj_id=2859&docType=pdf. Version from 05 March 2021
18. Aircraft Recycling –The Life and Times of an Aircraft (2021). https://airlines.iata.org/analysis/aircraft-recycling-the-life-and-times-of-an-aircraft. Version from 06 March 2021
19. Recycling Best Management Practice (BMP) for Management of Used Aircraft Parts and Assemblies (2021). https://afraassoc.wpengine.com/wp-content/uploads/2016/03/Recycling-BMP-Rev-4.pdf. Version from 06 March 2021

# Measures to Increase the Extra-Transport Effectiveness

Sergey Vakulenko[1], Peter Kurenkov[1(✉)], Grigoriy Deruzhinskiy[2], Alexander Ignatenko[3], Vladimir Nekhoroshkov[4], Alyona Aroshidze[4], Nikita Moisievich[5], and Irina Seryapova[6]

[1] Institute of Management and Digital Technologies, Russian University of Transport (RUT–MIIT), 9, Building 9, Obraztsova Street, Moscow 127994, Russia
post-iuit@bk.ru

[2] Faculty of Water Transport Operation and Navigation, Admiral F. F. Ushakov State Maritime University, 93, Lenin's Avenue, Novorossiysk 353918, Russia
evropa@bk.ru

[3] Marine Engineering Faculty, Admiral F. F. Ushakov State Maritime University, 93, Lenin's Avenue, Novorossiysk 353918, Russia

[4] Faculty International Business and Law, Siberian Transport University, 191, Dusi Kovalchuk Street, Novosibirsk 630049, Russia

[5] Directorate of Terminal and Warehouse Complexes, Administration of the Kuibyshev Railway, 2, Komsomolskaya square, Samara 443030, Russia

[6] Faculty of Economics, Logistics and Management, Samara State University of Railway Transport, 2V, Svobody Street, Samara 443008, Russia

**Abstract.** Railway transport is an integral part of the unified transport system and to be considered a key link within the Russian Federation economy. The general trend in the transport companies' development is to increase the transport services efficiency to benefit country's economy. The purpose of this scientific article is to identify aspects of the non-transport effect, considering the railway transport enterprises performance quality and efficiency. The analysis was performed to highlight the relationship of environmental and economic indicators in the activities of transport companies. There were efficiency calculation methods, and the environmental protection measures relevant to transport companies' operational processes described. The analysis of the JSC "Russian Railways" key performance indicators (KPIs) within 2017–2019 was carried out. The main directions of the organizational development and technical measures to improve the non-transport effect are presented.

**Keywords:** Railway transport · Economy · Development · Transport company · Service · Non-transport effect · Quality · Efficiency · Indirect impact

## 1 Introduction

The costs incurred by a transport company produce a result or effect both in the transport industry and beyond. Transport companies in Russia are successfully developing, the


G. Sierpiński (Ed.): TSTP 2021, LNNS 352, pp. 133–142, 2022.
https://doi.org/10.1007/978-3-030-91156-0_11

transport infrastructure is being improved, and the volume of freight and passenger traffic is increasing. At the same time, the problem of environmental pollution continues to be relevant and important for the country's transport companies [1].

The development of an ecologically efficient transport system aims to reduce the environmental degradation caused by transport-related noise and pollution [18]. Human activity affects ecosystem functions through the disruption of pattern and rate of matter as well as energy flow through ecosystems [17].

JSC "Russian Railways" is Russian state-owned integrated company that holds strategic positions in the Russian Federation transport complex, that, in turns, has a beneficial effect for all the aspects of social and economic development of the entire country, as well as for improving the life quality for all the citizens. Therefore, the non-transport effect and environmental protection aspects assessment has a high influence for the society as such and will reflect its importance for the entire country's population [2].

## 2 Materials and the Impact of Transport on Economic Development

The problems relevant to the non-transport effect assessing and measuring are highlighted in the scientific papers: P.V. Kurenkov, A.F. Kotlyarenko, N.P. Tereshina, A.P. Abramov, Yu.G. Sokolov, A.V. Ryshkov, A.V. Shobanov, I.M. Lavrov, and B.M. Lapidus, D.A. Macheret, M.F. Trikhunkov, I. Docherty, D. Banister. The non-transport effect and its change is reflecting the nature of the basics of transport operations and the growth of the socio-economic effect within urbanized areas and at the same time highlights the problem of its the negative impact on the environment [3, 19–22].

The transport impact for the social life and the development of other sectors of the economy can be assessed by the extra-transport effect. Benefits, associated gains or losses obtained in various spheres of the social-economic life and sectors of the national economy, as a result of the use of one or another type of transport and various transport technologies, while not reflected in the financial indicators of transport enterprises are reflected in the extra-transport effect [4].

High profit's role in enterprise development and promoting the interests of its owners and staff determine the need for effective and continuous process of development and management decisions on all major aspects of the formation, distribution, use and planning of business profits [23].

The goods delivery accelerating, for both direct and combined traffic, has a profound and multifaceted impact on both transport itself and other sectors of the economy. At the same time, a saving investments in transport sector is achieved, the capital assets renewal is accelerated, and the capital productivity is increased.

The freights outcome is the value of cargo delivered to the destination – i.e., to its "consumption point". It should be considered as one of the KPIs. Every transport operation outcome has also an indirect effect outside of transportation industry: for both the enterprises serviced by transportation and for the population using these services. And this value can be measured and compared as the cumulative effect resulting from the extra-transport effect associated with the costs of serviced production and in-transport, characterized by transportation costs.

The signs of efficiency characterizing the non-transport effect are: complete and timely (uninterrupted) satisfaction in transportation services matching the needs of the national economy and the population; high quality of transport services for the national economy and its population, characterized by the costs of enterprises associated with the receipt, storage and consumption of the arrived cargo, and the costs of dispatching finished products in terms of freight traffic; high speed of delivery, differentiated by economical basis; increasing the transportation safety level; reducing the need for low-skilled labor within the industry, agriculture, construction, trade, and associated auxiliary operations.

The signs of efficiency those reflects the intra-transport effect are: optimal distribution of traffic between modes of transport based on the transportation performance of work in the field of their most effective application, taking into account the quality of transport services; proportionality of the transport network elements development for different types of transport in economical regions and in interacting directions; coordinated operation of modes of transport along the entire route of cargo from the place of production to the place of consumption and passengers from the point of departure to the point of destination; low cost of transportation, subject to proper quality and other requirements; sufficient power and maneuverability of the system; traffic safety; the right combination of centralized and regional management of the transport system.

In the USSR, the method of complex analytical comparisons was used, but at present it cannot be applied due to the fragmentation of the transport modes management. The essence of this method is that not only the costs of wagon-hours are considered and compared, but also the costs of ship-hours, with the corresponding bringing them into a comparable form according to the cost estimate of ton-hours of wagons and ton-hours of ships, and the costs of ton-hours of cargo, costs in national currency (Russian rubles) or other currency for the entire route of cargo in mixed traffic.

The main purpose of this method is to choose an option that provides a reduction in time (car-hours, ship-hours, ton-hours), as well as in the costs. To achieve this goal, the main focus stays not on the purpose of the wagon or another vehicle, but on the purpose of the cargo being transported by vehicle.

The main problem in assessing the effectiveness of measures to accelerate the delivery of goods is that costs are carried out within individual links of transport chains, and the bulk of the effects are felt outside them. Therefore, the costs are quite real, they are quite simple to quantify. But the magnitude of the non-transport effect can be judged only tentatively: within a separate link of trade and transport chains, as a rule, only conditions are created for its appearance. The dispersion of the acceleration effect in time, space and in terms of its recipients, many "extraneous" factors affecting the dynamics of indicators, those could be used to judge the effectiveness of the activities, making this task as interesting as it is and difficult as well.

These circumstances, on the one hand, make it difficult to compare costs and benefits when justifying relevant projects, and on the other hand, they generate contradictions between private and public interests.

In connection with the creation of a competitive environment in all types of transport and at the points of their interaction, the economic interest of the subjects of transport hubs only is the results of their activities. It makes no sense to expect that commercial structures would perform activities or put plans for implementation with strong sectoral or national significance. Therefore, there is a need to redistribute the effect of accelerating the transportation process. The tool of redistribution depends on the degree of technological connectivity between the links within the transport chain. However, any system should be based on an assessment of the effect to be redistributed.

In addition, the need to consider the efficiency and quality of transport services necessarily arises, based on the interests of carriers, forwarders, cargo owners and other participants of the processes within transport hub.

Thy method proposed by V.A. Makeev in 2000–2001 is using an oblique matrix, similar to the matrix of indicators for the transport services quality, life and management of the region. The number of rows and columns in the matrix may vary depending on the participants within the transport hub and the quality indicators for the transport services in foreign trade in mixed traffic – that reflects the matrix versatility. This approach allows us to monitor the relevant indicators fully and visually, as well as it highlights the most significant indicators for carriers, forwarders, cargo owners and others within the transport hub, that contributes to an increase in the effectiveness of their strategies.

The reserves of the non-transport effect from equipping the territory with railways and highways and accelerating the delivery of goods and passengers can exceed the effect on transport by 8–10 times.

A deep understanding of the phenomena that arises during the acceleration of cargo delivery, their economic interpretation allows not only to increase the reliability of methods for assessing the effectiveness of such acceleration, but also create conditions for its growth.

The theory of determining transport and non-transport effects received another approach in the paper of Tereshina N.P., Abramov A.P. and Galaburda V.G, those proposed to take into account not only the effects, but also negative impact from transport activities associated with environmental pollution, alienation of lands, forest areas and other natural resources, as well as noise, vibration and other factors affecting humans' health, and traffic accidents.

## 3 Study of the Factors Related to the Railway Transport Negative Environment Impact

The degree of transport impact on the environment is characterized by the fact that an increase in the consumption of natural resources entails an increase in the number of pollutants entering the natural environment of the regions where transport companies are located, and as an outcome, the costs of environmental restoration to increase.

Factors of the negative impact of transport on the natural environment and their classification are shown in Fig. 1. Environmental pollution affects human health in a various ways and can practically affect all aspects of human's life.

According to the World Health Organization, the state of public health by 20–30% depends on the solution of environmental problems, air pollution, that is a growing threat to public health and the well-being of society [5].

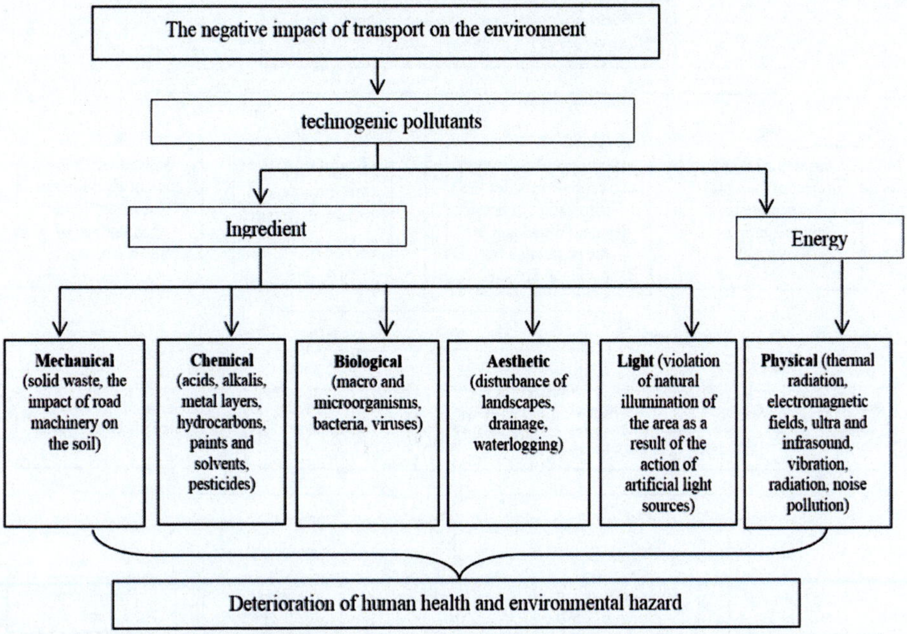


**Fig. 1.** Railway transport negative impact towards the environment: factors classification (proposed and developed by the authors)

## 4 Evaluation of the Effectiveness of Environmental Protection Measures

The social-ecological and economic efficiency of environmental protection is the ratio of total economic benefits and losses from environmental protection measures, including external environmental effects, and associated social and economic consequences. The structural "The logic and interconnection diagram" of social-ecological-economic assessment of the environmental protection measures effectiveness is presented in Fig. 2. Towards human health the risk assessment covers the assessment of the probability of hazardous pollution of air, water, soil, and its definition.

The environmental risk, its high degree, indicates the need for urgent environmental protection and rehabilitation and adaptation measures.

Transport consumes a large volume of natural resources and, as a result, uses significant amounts of fuel, pollutes the environment, requires the seizure of agricultural land for railways, roads and stationary structures, often causes injury and death of people [6]. The calculation of indicators for improving the economic, social, and environmental efficiency of transport companies is presented in Table 1 [7].

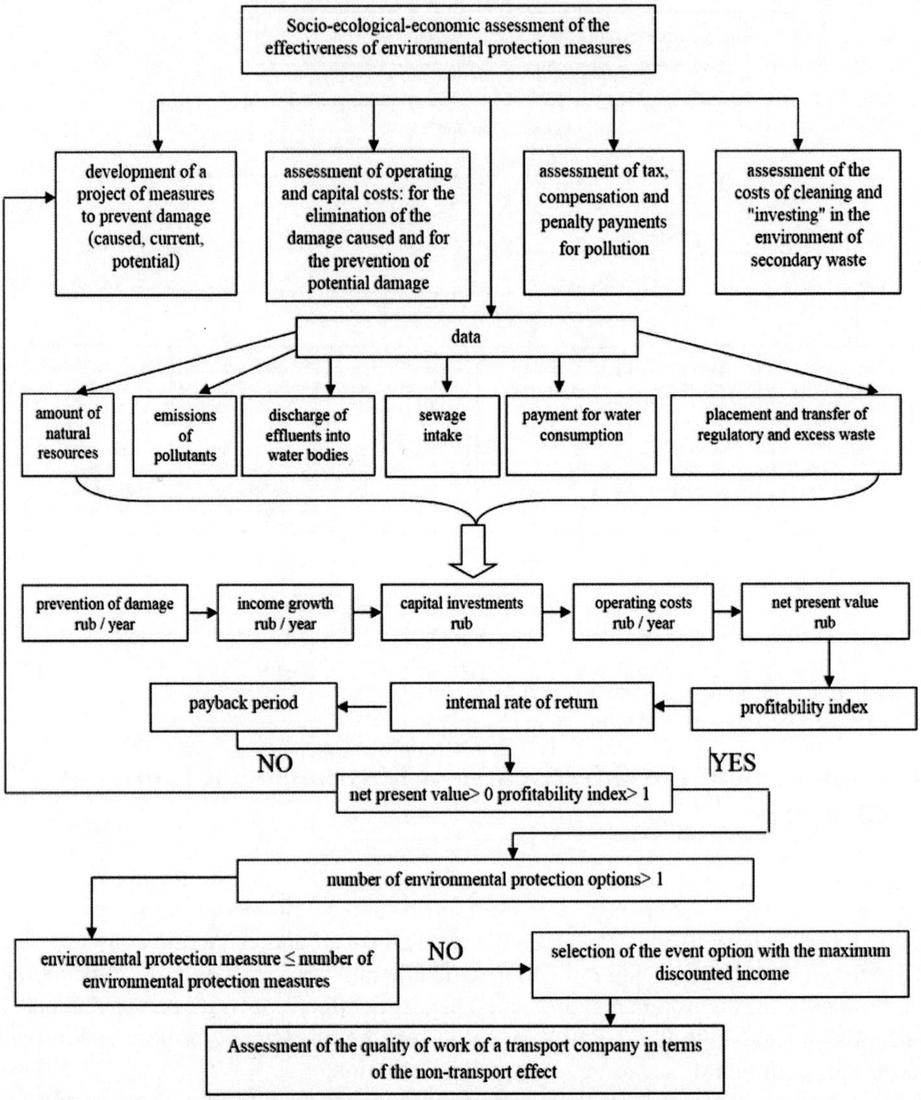


**Fig. 2.** Logic and interconnection diagram of socio-ecological-economic assessment of the effectiveness of environmental protection measures in the aspect of the non-transport effect (proposed and developed by the authors)

To improve the quality of transport services, a quality management system is being introduced into all management processes based on the concept of total quality management and ISO 9000 series standards. The purpose of the quality management system at railway transport enterprises is to ensure customer-oriented processes by continuously improving the quality of transport and logistics services to the level of satisfaction in the services provided in the field of freight and passenger transportation processes, improving traffic safety based on quality control of technological and business processes [8].

The increase in the population reduction, the depletion of natural and energy resources are the consequences of pollution of the biosphere with oil products, ions of heavy metals, surfactants [9].

Quality interacts with efficiency. Quality and efficiency are not of the same concepts. Quality and efficiency can vary in different directions. In most cases, an increase in quality leads to an increase in efficiency (Table 2).

Environmental measures are effective and expedient if, as a result of calculations, $Ep \geq 1$. At $0 \leq Ep \leq 1$, measures are not very effective, but if this is the only option to prevent an environmental catastrophe, then these measures make sense. If $Ep \leq 0$, then the activities are unprofitable and it is necessary to search for other solutions [10, 11].

**Table 1.** Indicators of improving the economic, social and environmental efficiency of transport companies.

| Efficiency | Calculation indicator | Symbols |
|---|---|---|
| Economic | $R_э = X_{1.1} + X_{1.2} + \ldots + X_{1.10}$ | $R_э$ – the total amount of points for economic performance indicators;<br>$X_n$ – the number of points awarded;<br>n – the ordinal number of the indicator |
| Environmental | $R_{np} = X_{2.1} + X_{2.2} + \ldots + X_{2.10}$ | $R_{np}$ – total score for environmental performance indicators;<br>Xn – the number of points awarded;<br>n – index number |
| Social | $R_c = X_{3.1} + X_{3.2} + \ldots + X_{3.10}$ | $R_c$ – the total score for social performance indicators;<br>$X_n$ – the number of points awarded;<br>n – index number |

The main directions for the organizational and technical measures development to improve the non-transport effect are shown in Fig. 3.

Unreliable transport operation or lack of convenient transport support may result in significant losses. This negative impact has to be considered as a loss in extra-transport effect (lost profit). This issue, in turn, can be resolved through the optimal development of the country's transport system.

In addition to the above, it should be noted that there are the following losses: due to inconsistency in the operation of the transport system elements, incomplete implementation of the increase in carrying capacity, inconsistency in the operation of the transport and trade systems, delay in settlements for transported goods between the seller and the buyer, as well as due to untimely payments between producers and consumers of transport services.

In order to prevent and eliminate the negative transport impact on the environment, it is necessary to widely apply organizational, technical, managerial and economic measures. Also, special attention should be paid to economic measures: optimization of taxation for enterprises in the aspect of non-transport effect, that involves environmentally friendly technologies application, the introduction of a scale of progressive fines for excess pollution, trade in quotas for environmental pollution, and other social, environmental, and environmental-economic projects developed [12, 13].

**Table 2.** Methods for calculating the effectiveness of environmental protection measures in the work of transport companies.

| Measures | Calculation indicator | Symbols |
|---|---|---|
| Benefits of environmental protection | $P = f(B, B_e, E, C_e)$ | $B$ - benefits;<br>$B_e$-benefits, taking into account environmental;<br>$E$ – expenses;<br>$C_e$ – costs of eliminating negative impacts, taking into account environmental components |
| Result of environmental activities (on a regional scale) | $P = (B + B_e) - (C + C_e)$ | $P$ - result of environmental protection; |
| | $B_e = \sum_{i=1}^{n} B_i$ | $B_i$ - benefits from specific $i$-th activities; |
| | $C_i = \sum_{i=1}^{n} C_i$ | $C_i$ - costs of specific $i$-th activities;<br>From 1 to $n$ - the number of events held |
| Result of environmental activities (at the 1 or several enterprises) emissions; consequences | $P_n = \sum_{j=1}^{n} (B_j - C_j)$ | $P_n$ - result, taking into account the specifics of enterprises;<br>$B_j$ – benefits from measures taken to prevent and scale eliminate the consequences of pollutant;<br>$C_j$ - costs of preventing and eliminating the of emissions of pollutants |

The non-transport effect from improving the provision of the territory with railways exceeds the effect on transport [14, 15].

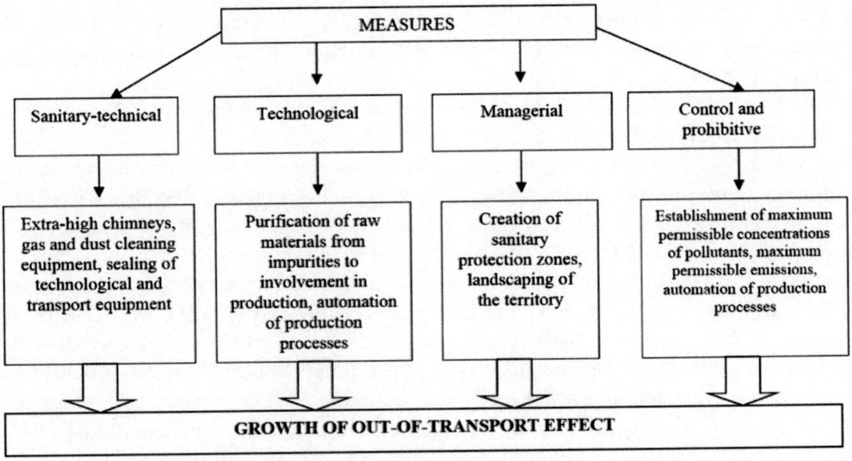


**Fig. 3.** Development of measures to increase the extra-transport effect

## 5 Conclusion

Thus, the basis of performance management is setting of the appropriate company's goal, due methods selection, and adequate analysis of internal and external performance results. The overall value of the non-transport effect may be assessed with the social-ecological and economic efficiency and its assessment has to be carried out in order to improve the quality and increase the efficiency of transport services. The implementation of specific measures within the transportation industry, such as building new railways, improving the quality of transport services, etc. would give a significant outer effect and these measures to be primarily focused on expanding supply and demand ratio.

Although the transport industry has obvious negative environmental impact related to its primary designation – to provide transport services and creating transport effect, in addition it also generates a non-transport effect for the economy. An increase in extra-transport efficiency can be achieved by assessing and systematizing environmental protection measures and counting on non-transport effect – sanitary, technological, managerial, control and regulatory measures.

## References

1. Katin, V.D.: Methods and devices for reducing emissions of harmful substances into the atmosphere from boilers at railway transport enterprises (Moscow: FGBOU "Educational and Methodological Center for Education in Railway Transport"), pp. 3–65 (2013)
2. Korablin, A.V., Vinogradov, S.V., Osipova, L.A., Sibryaev, K.O.: Protection of the Aquatic Environment from Transport Pollution. Kolos, Moscow (2010)
3. Sokolov, Yu., Galaburda, V.G., Lavrov, I.M., Anikeeva-Naumenko, L.O., Averyanova, O.A.: Extra-transport effect in quality management in railway transport. Transp. Bus. Russia **6**, 33–36 (2018)
4. Sokolov, Yu.I.: Methods for assessing the non-transport effect of improving the quality of transport services for cargo owners. Transp. Syst. Dev. Trends 543–545 (2016)

5. Sokolov, Yu.I., et al.: Management of demand for railway transportation and problems of market equilibrium (Moscow: Educational and Methodological Center for Education in Railway Transport) (2015)
6. Serdyukova, A.F., Barabanshchikov, D.A.: Influence of motor transport on the environment. Young Sci. **25**(211), 31–33 (2018)
7. Magomedov, S.S., Shlykov, E.E., Kuzina, M.A.: Comprehensive assessment of the quality and efficiency of the transport and construction industries. Postgraduate Readings 87–92 (2019)
8. Kuzina, E.L.: Ecological and economic assessment of the activity of the enterprise-nature user. Proc. Rostov State Univ. Civ. Eng. **15**, 113–120 (2011)
9. Kuzina, E.L.: Assessment of environmental and economic security of the country under the influence of transport factors economy. Entrepreneurship Environ. (EPIC) **1**, 162–169 (2012)
10. Tagiltseva, Y., Kuzina, E.L., Vasilenko, M.A., Drozdov, N.A.: Improving the mechanism for making managerial environmental decisions. Qual. Innov. Educ. **9**(148), 51–59 (2017)
11. Makeev, V.A.: Economic assessment of the feasibility of transferring and options for the construction of a railway line Current problems of transport development in Russia: strategic, regional, and technical (Rostov-on-Don RSUPS), pp. 387–388 (2004)
12. Sokolov, Y.I., Lavrov, I.M.: Methods of economic assessment of the quality of transport services for carriers in the conditions of the majority of participants in the transport process (Moscow: Golden Section) (2015)
13. Sokolov, Y.I.: Economics of the quality of Transport Services for Cargo Owners (Moscow: FGBOU " Training and Methodological Center for Education in Railway Transport) (2011)
14. Kurenkov, P.V., Zabennikov, V.S.: Holding companies in industry and transport: history, problems, prospect. Transport **9**, 23–28 (2014)
15. Kurenkov, P.V., Veselova, Y.V.: Mechanisms of state regulation of the Regional transport system development Transp. Sci. Technol. Manag. **7**, 8–15 (2006)
16. Ohl, C., Krauze, K., Grünbühel, C.: Towards an understanding of long-term ecosystem dynamics by merging socio-economic and environmental research. Ecol. Econ. **63**(2–3), 383–391 (2007). https://doi.org/10.1016/j.ecolecon.2007.03.014
17. Wasiak, M., et al.: Chosen aspects of simulation model to designing pro-ecological transport system. J. KONES Powertrain Transp. **21**(4), 525–532 (2014). https://doi.org/10.5604/12314005.1130519
18. Jacyna, M., et al.: Noise and environmental pollution from transport: decisive problems in developing ecologically efficient transport systems. J. Vibroengineering **19**(7), 5639–5655 (2017). https://doi.org/10.21595/jve.2017.19371
19. Docherty, I., Mackinnon, D.: Transport and Economic Development (2013). https://doi.org/10.4135/9781446247655.n13
20. Banister, D.: Transport and economic development: reviewing the evidence. Transp. Rev. **32**(1), 1–2 (2012). https://doi.org/10.1080/01441647.2011.603283
21. O'Brien, P.: Railways and the Economic Development of Western Europe, pp. 1830–1914 (1983). https://doi.org/10.1007/978-1-349-06324-6
22. Filip, N., Popa, C.: The role of transportation in economic development. "Mircea cel Batran" Naval Acad. Sci. Bull. **XVII**, 10–12 (2014)
23. Malekpour, S., Brown, R.R., de Haan, F.J.: Strategic planning of urban infrastructure for environmental sustainability: Understanding the past to intervene for the future // Cities. Victoria 3800, Australia: School of Social Sciences, Water for Liveability Centre and the Cooperative Research Centre for Water Sensitive Cities, Monash University (2015)

# The Impact of COVID-19 on Public Transport in Polish Cities on the Example of Wroclaw

Agnieszka A. Tubis(✉)

Faculty of Mechanical Engineering,
Wroclaw University of Science and Technology, Wroclaw, Poland
agnieszka.tubis@pwr.edu.pl

**Abstract.** The COVID-19 pandemic is an example of a "black swan", an unexpected event with a huge impact on the global economy. The introduced restrictions, aimed at reducing the level of virus emissions, influenced not only the economies of countries, but also the everyday life of their inhabitants. For this reason, it is currently one of the most up-to-date research topics of scientists around the world. One of the important areas of research is the impact of COVID-19 on the transport behaviour of the population. For this reason, the aim of the article is to present the results of the literature research on changes in the transport behaviours of residents during the COVID-19 pandemic and to characterize the changes in the functioning of the public transport system in one of the large cities in Poland. On this basis, regarding guidelines related to the future functioning of public transport in the city, conclusions will be formulated.

**Keywords:** COVID-19 · Transport behavior · Public transport

## 1 Introduction

The COVID-19 pandemic is an example of an unexpected event that has a huge impact on the global economy. Therefore, it meets the criteria of the phenomenon described by Taleb in [1] and can be considered as the so-called "black swan". COVID-19 has been also the most complex public health crisis of the century [2], which has significantly influenced the lives of all nations in the world. Each country has introduced different degrees of restrictions to prevent and control the spread of the virus. This restrictions significantly influenced peoples' lifestyles, social interactions, and economic conditions [3]. Fear of infection and perceived risk also influenced the behaviour related to traveling and choice of means of transport by the inhabitants [4], especially in large cities.

Some authors note that public transport (PT) might contribute to the disease's spread due to the enclosed spaces and the peoples' agglomeration [5]. The research described in [6] indicates that public transport systems are highly vulnerable to disease outbreaks. For this reason, during a pandemic, the first government guideline was to limit the mobility of inhabitants [7]. In order to counteract the development of the pandemic, the authorities of France and the United States decided to close PT [8, 9]. Many governments have introduced guidelines to PT users in order to respect social-distancing, facemasks in


G. Sierpiński (Ed.): TSTP 2021, LNNS 352, pp. 143–155, 2022.
https://doi.org/10.1007/978-3-030-91156-0_12

buses, trains and limited number of passengers [10, 11]. Tarasi et al. note, however, that the social distancing that epidemiologists encourage, that is to say, at least 2 m distance between individuals, is incompatible with public transit [5]. The number of passengers has dropped drastically in many countries, because of asking people to stay home or search for other transport alternatives when it is absolutely necessary [4]. In particular, the guidelines on "stay at home" have profoundly impacted the economic activities in many cities around the world which in turn have deeply changed people's travel behaviors [12]. An example may be the effects of the restrictions introduced in France, where there was a 65% reduction in the number of trips, especially in reducing work-related short-distance mobility, especially at peak Times [13]. Whereas in Singapore, mass rapid transit has reduced hit due to the COVID-19 by 80% [14].

The sharp decline in demand due to the perceived risk of infection in vehicles had negative consequences. One of the observed results is a drastic reduction in the current revenues from fees of collective transport operators [15]. It should also be emphasized that the drop in the use of public transport is also a significant issue for sustainable urban transport planners [16]. However, the currently observed situation may also have negative consequences in the future. The current limitations of the demand for public transport may be particularly dangerous for the future transport behaviour of inhabitants of large cities. If users switch from PT to private cars or resign from traveling at all, then after the pandemic is over, it may have a negative impact on PT's competitiveness due to changes in PT management in times of lower demand [17]. Meanwhile, the international association of public transport (UITP) considers the maintenance of high levels of service despite the reduction in travel demand to ensure safe distancing as one of the main challenges associated with resuming public transport operations [18].

The impact of the COVID-19 pandemic on public transport is the subject of numerous studies and assessment of the current situation (still during the pandemic). For a year now, a regular increase in the number of publications in this area can be observed. Their analysis is particularly important not only due to the observation of current transport behaviour, but also to predict future attitudes of residents. For this reason, the article aims to present the results of the conducted literature research on changes in the transport behaviour of residents during the COVID-19 pandemic and to characterize changes in the functioning of the public transport system in one of the large cities in Poland. On this basis, conclusions will be formulated regarding guidelines related to the future functioning of public transport in the city. The structure of the article assumes the presentation of the adopted research methodology in point 2. Next, the results of the literature review and evaluation of the functioning of the selected public transport system will be presented in point 3. On this basis, conclusions and predictions about the future functioning of public transport in the city will be formulated, as presented in point 4.

## 2 Methodology and Research Area

The assessment of the impact of the COVID-19 pandemic on the functioning of the public transport system in large cities is only in the first phase of the analysis. In many countries in Europe, including Poland, there are still restrictions on the movement of residents using public transport and forced remote work (home office). This significantly determines the city traffic in particular cities. For this reason, at present, only the direct effects of restrictions that are introduced at the state and municipal level can be observed. However, many researchers try to make attempts to forecast the long-term impact of this event on the transport behaviour of residents of large cities around the world. The research presented in this article aims to analyse the current functioning of public transport during the COVID-19 pandemic and to try to determine the factors influencing the future transport behaviour of inhabitants of large cities in Poland. Therefore, the conducted includes three steps of the procedure presented in Fig. 1.

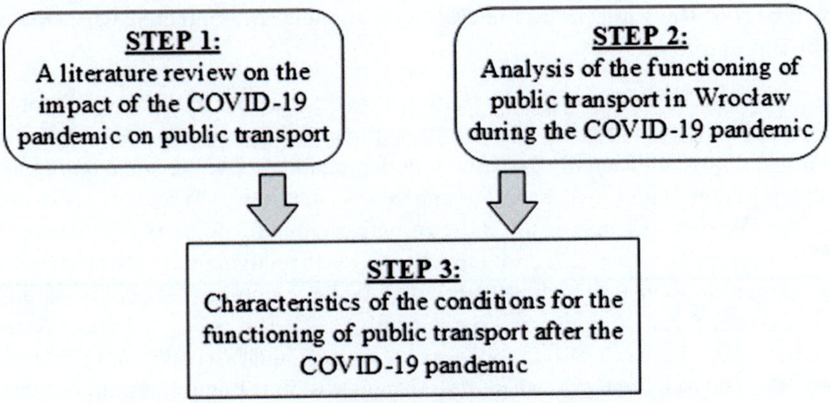


**Fig. 1.** Research procedure

The research methodology of the first two areas is presented in Sects. 2.1 and 2.2. While, a cause-and-effect analysis based on the results obtained in the first two research procedures was used to formulate conclusions regarding the conditions of public transport in the future, post-COVID.

## 2.1 Literature Review

The aim of the literature review is to identify scientific publications that present the impact of the COVID-19 pandemic on the transport behaviour of residents or describe the functioning of public transport during the pandemic. The publications were searched for in available journals databases, which they publish in the form of "open access" and in the form of subscriptions available by Wroclaw University of Science and Technology. The following set of keywords was used for the search: *urban transport + COVID-19 + pandemic*. These words could be used by the authors in any part of the publication.

As a result of the procedure (as of 15/04/2021), 113 items were returned, which were published in peer-reviewed journals. Among these items, the following were distinguished: 107 articles, 5 conference materials, 1 review. All found items were published in the years 2020–2021. However, it should be noted that due to the timelines of this research area, the potential number of articles on the studied issue increases every day.

Due to the purpose and scope of the conducted research, not all publications were a source of information for the conducted analysis. Some of the articles obtained as part of the search concerned general transport behaviour, and thus also long-distance travel. For this reason, the collection of publications was limited only to those directly related to urban transport. From the remaining group of publications, publications concerning the impact of limited traffic in cities on air quality and urban planning in the post-COVID period were then eliminated. Ultimately, only those articles were selected for the analysis that directly related to the impact of the COVID-19 pandemic on the functioning of public transport in cities and changes in transport behaviour of residents regarding their use of public transport. For this reason, 26 items were used for the detailed literature review.

### 2.2 Analysis of the Functioning of Public Transport in a Selected City During the Pandemic

Research on the functioning of public transport was carried out in Wroclaw. Wroclaw is one of the largest cities in Poland, with a current population of around 700,000. Transport traffic in the city is additionally generated by the inhabitants of the surrounding towns, who regularly commute to work, and young people studying at Wroclaw's schools and universities. Wroclaw is one of the most attractive cities for tourists in Poland. In the *Best European Destinations 2018* ranking, the city was recognized as the best European tourist destination to visit, which is reflected in the large number of tourists visiting the capital of Lower Silesia (about 5 million annually) [19]. The public transport system in the city includes a network of bus connections (normal and fast) and tram connections. It is also planned to increase the share of transport within the agglomeration railway by several years - in particular for people commuting to Wroclaw from nearby towns.

- The analysis of the functioning of public transport in the city concerned three research areas:
- Analysis of documentation and recommendations of the carrier MPK Sp. z o.o. due to the introduction of a pandemic state.
- Direct observation of the pandemic situation in Poland and in Wroclaw - analysis of press releases, restrictions introduced at the central and local level, and the behaviour of residents.
- A survey conducted among the residents of Wroclaw as part of the master's thesis prepared by Izabela Bohdanowicz under the supervision of dr hab. Agnieszka Tubis [20].

The survey was conducted from November 24, 2020, to March 16, 2021, in the form of an interactive survey using a Google form. This is the period of the 2nd and 3rd waves of the pandemic. The questionnaire form contained open, semi-open and closed questions, of the conjunctive type allowing the selection of several possible answers, and of the disjunctive type, in which the respondent may choose only one of the given options. A total of 570 respondents participated in the study, of which women constituted 56.9% of the respondents, and men 43.1%.

# 3 Results

## 3.1 Literature Review

After the selection, 26 articles were chosen for detailed analysis. All publications are from 2020–2021 due to the period when the COVID-19 pandemic began (announcement of the state of the pandemic - March 2020). As shown in Fig. 2, most of the analysed articles were published in three magazines: *Transportation Research Interdisciplinary Perspectives, Sustainability* and *Transportation letters.* Individual articles also came from magazines such as:

- *Transportation,*
- *Travel Behaviour and Society*
- *International Journal of Transportation Science and Technology*
- *Global Transitions*
- *The Lancet Digital Health*
- *Transport Reviews*
- *Urban Transport of China*
- *Journal of Public Transportation.*

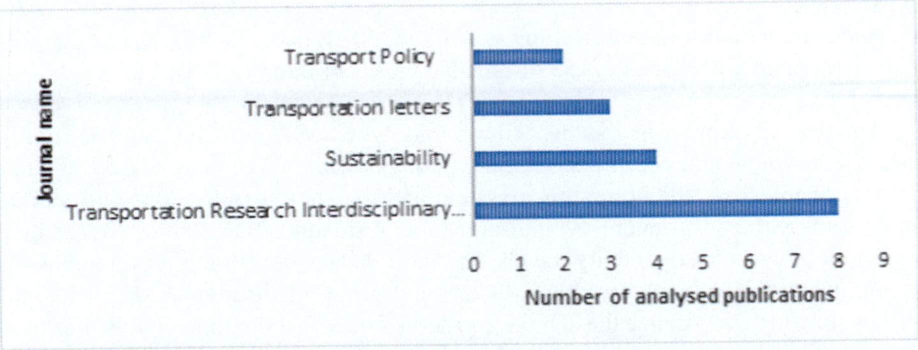


**Fig. 2.** Number of analysed publications appearing in individual journals

The content-related analysis of the articles made it possible to distinguish 3 research areas, which are presented in Fig. 3. The publications were classified to each research area by the purpose of the article defined by the authors. As can be seen in Fig. 3, some items concerned simultaneously two issues related to research into the COVID-19 pandemic.

Currently, the most numerous group are publications on changes in transport behaviour (19 analysed articles). The conducted research was focused primarily on the analysis of changes in the behaviour of residents caused by the restrictions announced by the government, the introduction of home-office and the fear of contracting the virus. The analysis included:

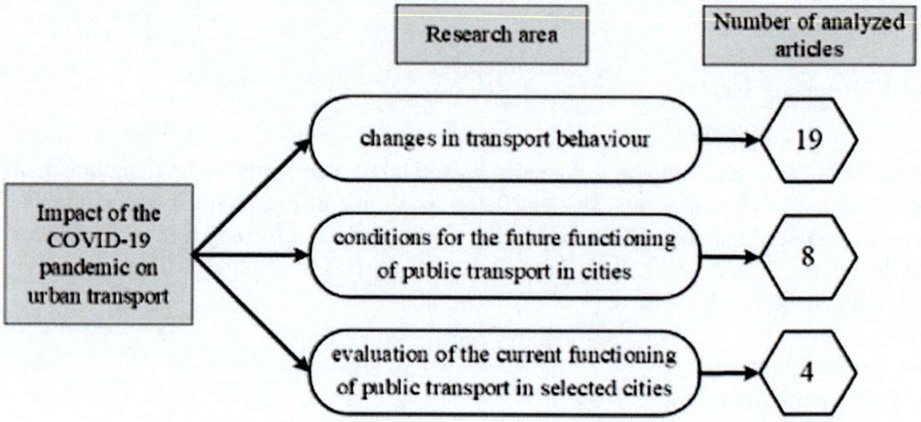


**Fig. 3.** Highlighted research areas described in articles on the impact of COVID-19 on public transport

- assessment of the sense of safety of public transport users [21–25];
- assessment of the impact of demographic and socioeconomic factors on transport behaviour during the pandemic [13, 26–28];
- identification of directions of changes in the transport behaviour of residents [2, 5, 12, 14, 29];
- characteristics concerning travelling to work [3, 30–32];
- assessment of willingness to use sustainable means of transport [16]

An analysis of published studies shows that the COVID-19 pandemic has significantly reduced the number of people using public transport. The main reasons include the fear of infection, the home-office system and the current restrictions on mobility introduced by the government. At the same time, residents of large cities choose their own car or bicycle for their daily travels. For short distances, walking is preferable.

An important area of research is the second group of distinguished articles, the purpose of which is to define the conditions for the future functioning of public transport in cities. Authors of these publications, based on the inference process, try to formulate guidelines for the design and implementation of solutions aimed at the development of public transport in the post-pandemic period [12, 15, 33–35]. Some articles also analyse critical issues related to the use of public transport to understand the factors influencing future transport behaviour [25, 36, 37].

The third thematic group were publications on the evaluation of the current functioning of public transport in selected cities. The main subject of the assessment was the implemented solutions limiting the risk of infection, including, in particular, restrictions on the number of passengers in public transport vehicles [14, 26, 32]. The analyses also included an assessment of the degree of compliance with the introduced restrictions by the inhabitants of selected regions [38].

### 3.2 Functioning of the Public Transport System in Wroclaw

Wroclaw, like other cities, introduced restrictions on the use of public transport in a short time from the moment of the first detection of COVID-19 infection in Poland. In addition to the nationwide recommendation on limiting mobility, maintaining social distance and the obligation to cover a mouth and nose (the requirement to wear masks), Wroclaw also introduced restrictions on the permissible number of people travelling by particular means of transport. This restriction applied to both buses and trams. Detailed guidelines regarding the permissible number of passengers in some types of vehicles used in public transport in Wroclaw are presented in the Table 1. The table also shows the level of permissible load of vehicles indicated by the manufacturer. It is worth noting, however, that in the period before the pandemic, during rush hours, selected lines recorded situations of overcrowding in the number of passengers, which even exceeded the acceptable standards.

**Table 1.** The permissible number of passengers according to the vehicle manufacturer's guidelines and in force during the pandemic.

| Vehicle brand | Manufacturer's guidelines | Permissible number of passengers during pandemic |
|---|---|---|
| **BUSES** | | |
| Mercedes-Benz Solo | 102 | 51 |
| Volvo | 176 | 88 |
| Solaris | 146 | 73 |
| **TRAMS** | | |
| Moderus Beta | 214 | 107 |
| Skoda 19T | 202 | 101 |
| Pesa Twist | 222 | 111 |

The introduced restrictions on the number of passengers in public transport were not negatively felt by the inhabitants of the city. This was mainly since many of them gave up this form of moving around the city. This was confirmed by the research [20], which indicated that 67.1% of the respondents reduced the number of their journeys in public transport. As shown in the Fig. 4 most of the decisions were made by people who used this form of travel every day before the pandemic.

In many organizations located in Wrocław, remote work or work in a rotational system has been introduced, in which employee teams work stationary and alternately (individual teams work on selected days of the week in the home office system). At the same time, all universities, and then primary and secondary schools, switched to the remote learning mode, which significantly reduced the number of users of public transport. This is confirmed by studies in which the respondents indicated "fear of being infected" (236 people) and "obligatory distance work/learning" (167 people) as the two main reasons for resignation or limiting the frequency of using public transport.

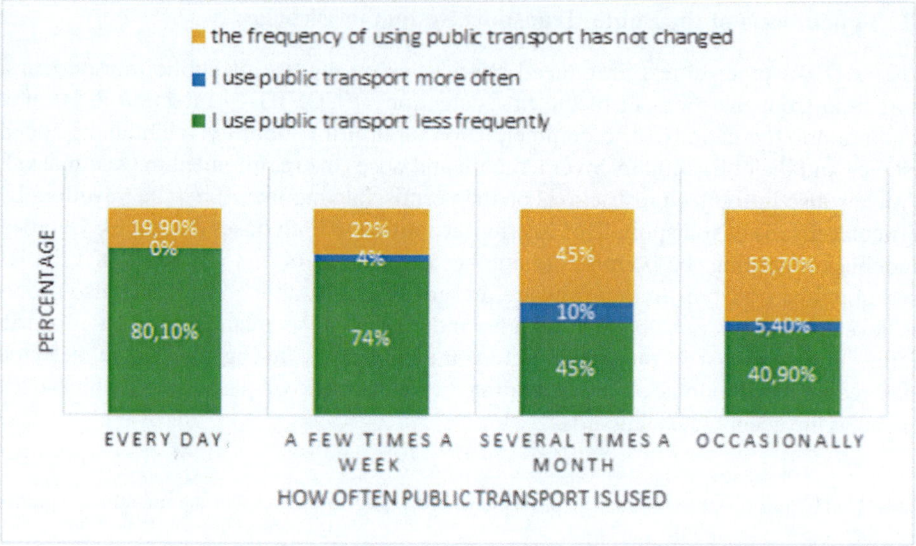


**Fig. 4.** Share of changing transport behaviour during a pandemic COVID-19 (Source: [20])

In addition to the nationwide recommendations on wearing masks and maintaining a safe distance, as well as introducing passenger limits in public transport vehicles, the carrier MPK Sp. z o. o. introduced additional crisis management measures:

- Designation of a buffer zone for the driver – R1
- Automatic opening of all doors in vehicles - elimination of the so-called 'hot buttons' – R2.
- Recommendation to occupy every second seat – R3.
- Launching special transports for 'white personnel' – R4.
- Messages reminding of the applicable safety rules – R5.
- Ozonation of vehicles – R6.

All the above activities were assessed very high by the inhabitants. Detailed assessments of each of the activities are shown in the Fig. 5.

At the same time, however, the respondents admitted in the research that they do not adapt to all the restrictions introduced in public transport vehicles. The most frequently ignored recommendations during the journey include not respecting the passenger limit in the vehicle and keeping a safe distance. Most of the respondents (over 50%) indicated in both these cases that they permanently or sometimes ignore these recommendations [20].

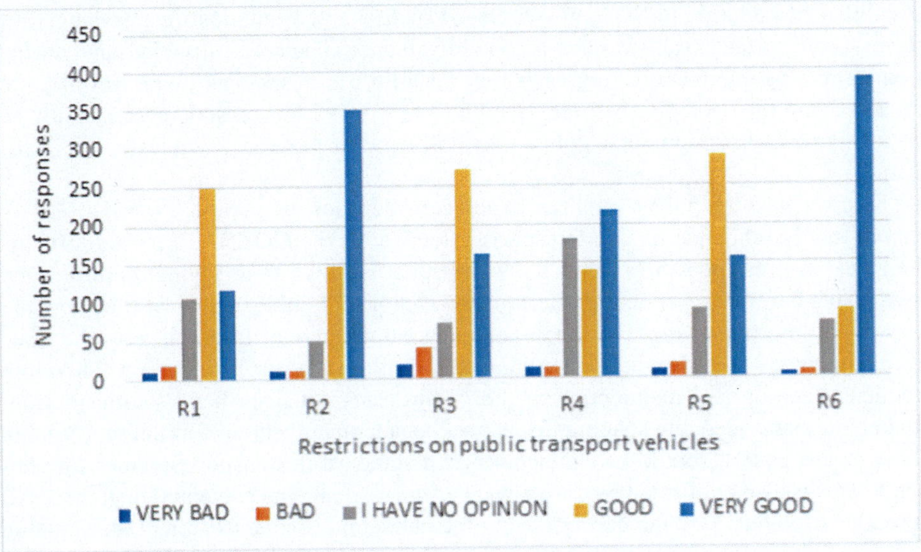


**Fig. 5.** Assessment by residents of the most important restrictions introduced in public transport vehicles during the pandemic in Wroclaw (Source: [20])

## 4 Conclusions and Discussion

Based on a review of the literature, analyses of the functioning of public transport during the pandemic, as well as observations of the behaviour of residents, an attempt was made to determine the future shape of the functioning of public transport in cities.

The literature review clearly shows that the COVID-19 pandemic has a significant influence on travel patterns. The reduction in traffic volume impacted by this pandemic has been seen in cities all over the world [39]. The effect of self-isolation, social distancing, and the shuttering of shops, restaurants and offices from the pandemic could be seen in the traffic data [40]. Unfortunately, the studies presented by some authors prove that transit ridership may not be fully recovered even after the COVID-19 restrictions are lifted [26, 41]. Additionally, there is a risk that if the public transportation sector is perceived as poorly transitioning to post-pandemic conditions, that viewing public transportation as unhealthy will gain ground and might be sustained [36].

As noted by Ozbilen et al. [24] transportation mode choice is associated with a number of factors, namely socio-demographics, built environment characteristics, travel attitudes, intrinsic values of mobility, and risk perceptions. The latter two are particularly important during the COVD-19 era, as they determine the mode choice in emergency conditions. For this reason, fear of infection and perceived risk could also affect travel behaviours and mode choices in the pandemic period, but also in future transport behaviours [3]. The results of the research in Wroclaw, which were presented in point 3 of this article, also confirm that the fear of infection is one of the main reasons for giving up using public transport vehicles.

Avoiding public transport due to the fear of contamination is understandable. First of all, during the pandemic, public transport vehicles were stigmatized for increasing the possibility of infection [42]. Confined, closed spaces have been identified as dangerous places, prone to faster transmission of the virus. It should be noted, however, that during a pandemic, people perceive a greater risk for all types of trip and avoid travelling to places, where they perceive medium to high risk [43]. For this reason, efforts should be made to strongly communicate public transport benefits to society and restore citizens' trust.

Future transport policy requires better consideration of people's travel needs to restore their confidence in safe transport systems after the COVID-19 period [30]. As the research presented in [3] shows, factors such as travel time saving, comfort, and cost, which before the pandemic determined the choice of public means of transport, are now much less important. The inhabitants now attach greater importance to the issue of maintaining social distance, cleanliness and limiting the risk of infection. Therefore frequent cleaning and disinfection of the public transport fleet, hand sanitisers, floor stickers to mark adequate spacing [5], as well as adequate vehicle ventilation [36], are some measures that should become currently a standard in transport services for residents. This is also confirmed by the answers of the respondents presented in point 3. All activities introduced by the carrier during the pandemic (ozone treatment, designation of buffer zones, informing about the need to keep distance and cover the face) were positively assessed by residents.

At the same time, the relationship between mobility, urban space and health needs to be rethought immediately. It is necessary to transform urban environments after the pandemic by means of new sustainable transport models [44, 45]. De Vos [25] emphasizes that due to the pandemic, people will prefer to use active modes or cars over public transport. This is mainly due to the fact that active modes, motorcycle, and personal car are perceived by the participants as the least risky urban transportation modes during the pandemic [22]. According to a report by Lime Micro-mobility, people are more likely to choose more flexible, short-distance, greener, and community-focused modes, such as bicycles, e-scooters, and e-bikes [41]. However, in order to reduce the number of people travelling in the city by car, decision-makers will have to improve the condition of the city's infrastructure in terms of the availability of bicycle paths and the required parking spaces for bicycles and scooters. As [3] notes, the lack of a properly developed bicycle infrastructure is one of the reasons for not using this individual means of transport, indicated by the inhabitants. For this reason, many local administrators are taking action in their cities to make the bicycle the most efficient and safest means of transport for post-pandemic mobility in urban areas. An example of such solutions is the increase in the number of bike lanes (permanent and temporary) and sharing mobility programs in most Italian cities [35].

## 5 Summary

The article presents an overview of publications from the last 2 years concerning the COVID-19 pandemic. In the first analysed period, the researchers focused on the analysis of the impact of the pandemic on the behaviour of residents and assessed its impact on the functioning of the system. Currently, more and more publications focus on the post-COVID of the future and the actions that need to be taken to shape public transport in the conditions of the new reality.

As noted in [42], public transport is strongly interlinked with many other challenges (climate, health, social inclusion, road safety, etc.). Without prioritizing the development of these services and their adaptation to the new post-COVID reality, cities and their inhabitants will take a step back in their sustainable transport development. At the same time, as emphasized by Barbarossa [35], the ambition of the post-pandemic city should be not to return to 'normal' but to build a better, more sustainable and resilient society. 'Post-COVID normal' should be about new concrete measures that are put in place for recovery that support the required transition to a more sustainable, low-carbon, inclusive and healthier economy for people and the planet [35].

The research presented in this article will be continued. Currently, a massive vaccination campaign against COVID-19 is underway in Poland. More and more organizations and enterprises expect employees to return to regular stationary work. Reducing the restrictions also applies to the return of students to schools and universities. This means the necessity to constantly monitor the transport behaviour of residents, the functioning of the public transport system and to conduct research on the preferred means of transport again in the post-pandemic period.

## References


1. Taleb, N.N.: The Black Swan : The Impact of the Highly Improbable. Random House, New York (2007)
2. Chan, H.-Y., Chen, A., Ma, W., Sze, N.-N., Liu, X.: COVID-19, community response, public policy, and travel patterns: a tale of Hong Kong. Transp. Policy **106**, 173–184 (2021). https://doi.org/10.1016/j.tranpol.2021.04.002
3. Abdullah, M., Dias, C., Muley, D., Shahin, M.: Exploring the impacts of COVID-19 on travel behavior and mode preferences. Transp. Res. Interdiscip. Perspect. **8**, 100255 (2020). https://doi.org/10.1016/j.trip.2020.100255
4. de Haas, M., Faber, R., Hamersm, M.: HowCOVID-19 and the Dutch 'intelligent lock-down'change activities, work and travel behaviour: evidence from longitudinal data in the Netherlands. Transport. Res. Interdiscip. Perspect. **6**, 100150 (2020). https://doi.org/10.1016/j.trip.2020.100150
5. Tarasi, D., Daras, T., Tournaki, S., Tsoutsos, T.: Transportation in the Mediterranean during the COVID-19 pandemic era. Glob. Transitions **3**, 55–71 (2020). https://doi.org/10.1016/j.glt.2020.12.003
6. Edelson, P., Phypers, M.: TB transmission on public transportation: a review of published studies and recommendations for contact tracing. Trav. Med. Infect. Dis. **9**(1), 27–31 (2011)
7. Muley, D., Shahin, M., Dias, C., Abdullah, M.: Role of transport during outbreak of infectious diseases: evidence from the past. Sustainability **12**, 7367 (2020). https://doi.org/10.3390/su12187367
8. Guan, L., Prieur, C., Zhang, L., Prieur, C., Georges, D., Bellemain, P.: Transport effect of COVID-19 pandemic in France. MedRxiv (2020). https://doi.org/10.1101/2020.07
9. Lyu, W., Wehby, G.L.: Community use of facemasks and COVID-19: evidence from a natural experiment of state mandates in the US. Health Aff. **39**(8), 1419–1425 (2020). https://doi.org/10.1377/hlthaff.2020.00818
10. Hadjidemetriou, G.M., Sasidharan, M., Kouyialis, G., Parlikad, A.K.: The impact of government measures and human mobility trend on COVID-19 related deaths in the UK. Transport. Res. Interdiscip. Perspect. **6**, 100167 (2020). https://doi.org/10.1016/j.trip.2020.100167

11. Zhang, L., Yang, H., Wang, K., Zhan, Y., Bian, L.: Measuring imported case risk of COVID-19 from inbound international flights—A case study on China. J. Air Transp. Manag. **89**, 101918 (2020). https://doi.org/10.1016/j.jairtraman.2020.101918
12. Cho, S.-H., Park, H.-C.: Exploring the behaviour change of crowding impedance on public transit due to COVID-19 pandemic: before and after comparison. Transp. Lett. **36**, 367 (2021). https://doi.org/10.1080/19427867.2021.1897937
13. Pullano, G., Valdano, E., Scarpa, N., Rubrichi, S., Colizza, V.: Evaluating the effect of demographic factors, socioeconomic factors, and risk aversion on mobility during the COVID-19 epidemic in France under lockdown: a population-based study. Lancet Digit. Heal. **2**, 638–649 (2020). https://doi.org/10.1016/S2589-7500(20)30243-0
14. Gkiotsalitis, K., Cats, O.: Public transport planning adaption under the COVID-19 pandemic crisis: literature review of research needs and directions. Transp. Rev. **41**(3), 374–392 (2020). https://doi.org/10.1080/01441647.2020.1857886
15. Hörcher, D., Singh, R., Graham, D.J.: Social distancing in public transport: mobilising new technologies for demand management under the Covid-19 crisis. Transportation **2020**, 1–30 (2021). https://doi.org/10.1007/s11116-021-10192-6
16. Campisi, T., Basbas, S., Skoufas, A., Akgün, N., Ticali, D., Tesoriere, G.: The impact of COVID-19 pandemic on the resilience of sustainable mobility in Sicily. Sustainability **12**(21), 8829 (2020). https://doi.org/10.3390/su12218829
17. El Ouadi, J., Malhene, N., Benhadou, S., Medromi, H.: Shared public transport within a physical internet framework: reviews, conceptualization and expected challenges under COVID-19 pandemic. IATSS Res. (2021). https://doi.org/10.1016/j.iatssr.2021.03.001
18. UITP. COVID-19 Pandemic - Resuming public transport services post-lockown (2020). https://www.uitp.org/sites/default/files/cck-focus-papers-files/Knowledge_Brief_-_Covid19_0.pdf
19. https://pl.wikipedia.org/wiki/Wroclaw. Accessed 15 Apr 2021
20. Bohdanowicz, I.: The impact of the COVID-19 pandemic on the assessment of the quality of public transport by the inhabitants of Wrocław. Master thesis. Supervisor: dr hab. Agnieszka Tubis, Wroclaw (2021)
21. Przybylowski, A., Stelmak, S., Suchanek, M.: Pandemic-Public transport users in Gdansk case study. Sustainability **13**(1), 364 (2021). https://doi.org/10.3390/su13010364
22. Dingil, A.E., Esztergár-Kiss, D.: The influence of the Covid-19 pandemic on mobility patterns: the first wave's results Ali Enes Dingil & Domokos Esztergár-Kiss the influence of the Covid-19 pandemic on mobility patterns: the first wave's results. Transp. Lett. **13**, 434 (2021). https://doi.org/10.1080/19427867.2021.1901011
23. Jenelius, E., Cebecauer, M.: Impacts of COVID-19 on public transport ridership in Sweden: analysis of ticket validations, sales and passenger counts. Transp. Res. Interdiscip. Perspect. **8**, 100424 (2021). https://doi.org/10.1016/j.trip.2020.100242
24. Ozbilen, B., Slagle, K.M., Akar, G.: Perceived risk of infection while traveling during the COVID-19 pandemic: insights from Columbus, OH. Transp. Res. Interdiscip. Perspect. **10**, 100326 (2021). https://doi.org/10.1016/j.trip.2021.100326
25. De Vos, J.: The effect of COVID-19 and subsequent social distancing on travel behavior. Transp. Res. Interdiscip. Perspect. **5**, 100121 (2020)
26. Wang, D., He, B.Y., Gao, J., Chow, J.Y.J., Ozbay, K., Iye, R.S.: Impact of COVID-19 behavioral inertia on reopening strategies for New York City transit. Int. J. Transp. Sci. Technol. **10**, 197 (2021). https://doi.org/10.1016/j.ijtst.2021.01.003
27. Bhaduri, E., Manoj, B.S., Wadud, Z., Goswami, A.K., Choudhury, C.F.: Modelling the effects of COVID-19 on travel mode choice behaviour in India. Transp. Res. Interdiscip Perspect. **8**, 100273 (2020). https://doi.org/10.1016/j.trip.2020.100273

28. Habib, K.N., et al.: Assessing the impacts of COVID-19 on urban passenger travel demand in the greater Toronto area: description of a multi-pronged and multi-staged study with initial results. Transp. Lett. **13**, 353 (2021). https://doi.org/10.1080/19427867.2021.1899579
29. Li, A., Zhao, P., He, H., Axhausen, K.W.: Understanding the variations of micro-mobility behavior before and during COVID-19 pandemic period. ETH Zurich Research Collection (2020)
30. Pawar, D.S., Yadav, A.K., Choudhary, P., Velaga, N.R.: Modelling work- and non-work-based trip patterns during transition to lockdown period of COVID-19 pandemic in India. Travel Behav. Soc. **24**, 46–56 (2021). https://doi.org/10.1016/j.tbs.2021.02.002
31. Beck, M.J., Hensher, D.A.: Insights into the impact of COVID-19 on household travel and activities in Australia – The early days of easing restrictions. Transp. Policy **99**, 95–119 (2020). https://doi.org/10.1016/j.tranpol.2020.08.004
32. Sun, S., Wong, Y.D., Liu, X., Rau, A.: Exploration of an integrated automated public transportation system. Transp. Res. Interdiscip. Perspect. **8**, 100275 (2020). https://doi.org/10.1016/j.trip.2020.100275
33. Combs, T.S., Pardo, C.F.: Shifting streets COVID-19 mobility data: Findings from a global dataset and a research agenda for transport planning and policy. Transp. Res. Interdiscip. Perspect. **9**, 100322 (2021). https://doi.org/10.1016/j.trip.2021.100322
34. Shuang, L., Yu, Z.: Urban Transportation system planning under the COVID-19 pandemic in Beijing. Urban Transp. China **03**, 28–32 (2020)
35. Barbarossa, L.: The post pandemic city: challenges and opportunities for a non-motorized urban environment. An overview of Italian cases. Sustainability **12**(17), 7172 (2020). https://doi.org/10.3390/su12177172
36. Tirachini, A., Cats, O.: COVID-19 and public transportation: current assessment, prospects, and research needs. J Public Transp. **22**(1), 1–34 (2020). https://doi.org/10.5038/2375-0901.22.1.1
37. González-Sánchez, G., Olmo-Sánchez, I., Maeso-González, E.: Challenges and strategies for Post-COVID-19 gender equity and sustainable mobility. Sustainability **13**(5), 2510 (2021). https://doi.org/10.3390/su13052510
38. Dzisi, E.K.J., Dei, O.A.: Adherence to social distancing and wearing of masks within public transportation during the COVID 19 pandemic. Transp. Res. Interdiscip. Perspect. **7**, 100191 (2020). https://doi.org/10.1016/j.trip.2020.100191
39. Tian, X., An, C., Chen, Z., Tian, Z.: Assessing the impact of COVID-19 pandemic on urban transportation and air quality in Canada. Sci. Total Environ. **765**, 144270 (2021). https://doi.org/10.1016/j.scitotenv.2020.144270
40. Parr, S., Wolshon, B., Renne, J., Murray-Tuite, P., Kim, K.: Traffic impacts of the COVID-19 pandemic: statewide analysis of social separation and activity restriction. Nat. Hazard. Rev. **21**(3), 04020025 (2020)
41. Thigpen, C.: Rethinking Travel in the era of COVID-19: Survey Findings and Implication for Urban Transportation (2020). https://www.li.me/second-street/rethinkingtravel-in-the-era-of-covid-19-new-report-shows-global-transportation-trendssupport-for-micromobility
42. UITP: Public transport in COVID – safe (2020). https://cms.uitp.org/wp/wp-content/uploads/2020/10/Policy-Brief-PTisCOVID-Safe.pdf
43. Hotle, S., Murray-Tuite, P., Singh, K.: Influenza risk perception and travel-related health protection behavior in the US: Insights for the aftermath of the COVID-19 outbreak. Transp. Res. Interdiscip. Perspect. **5**, 100127 (2020). https://doi.org/10.1016/j.trip.2020.100127
44. Bucsky, P.: Modal share changes due to COVID-19: The case of Budapest. Transp. Res. Interdiscip. Perspect. **8**, 100141 (2020)
45. Batty, M.: The Coronavirus crisis: what will the post-pandemic city look like? Environ. Plan. B Urban Anal. City Sci. **47**, 547–552 (2020)

# Author Index





G. Sierpiński (Ed.): TSTP 2021, LNNS 352, p. 157, 2022.
https://doi.org/10.1007/978-3-030-91156-0